Museums, Art and Inclusion in a Climate Emergency

Museums, Art and Inclusion in a Climate Emergency considers the impact of the Anthropocene on history and memory, approaches to objects and agency and the incommensurability of western and Indigenous ontologies.

Drawing on Indigenous knowledge, humanities and museological literature, continental philosophy, contemporary art and popular culture, Baker acknowledges the autonomous agency of geological forms, including soils, minerals and fossil fuels. Demonstrating that this has implications for an expanded idea of an 'inclusive' museum and its relationship to entities beyond 'life' and living species, the book argues that the 'inclusion' paradigm needs to include nonlife actors. Gesturing to a geontological 'turn' through developing notions of geo-inclusion, the mineralhuman and approaches to object agency that connect with Aboriginal 'heritage', Baker exposes the ongoing destruction of Country by mining interests in Western Australia and elsewhere. By addressing the need for urgent change through the artifice of the museum, the book identifies an expanded approach to inclusion beyond the limits imposed by the politics of identity.

Museums, Art and Inclusion in a Climate Emergency theorises the potential of an expanded idea of the museum and will be of interest to scholars and students engaged in the study of museums and heritage, environmental humanities and geo-humanities, ecological art history and contemporary art.

Janice Baker is an independent curator, writer and scholar based in Western Australia. Her work facilitates creative research projects and programmes that support artists and environmental art history. Her next book is a cultural study of seaweed through the lens of art, museums and climate change.

Museums, Art and Inclusion in a Climate Emergency

Janice Baker

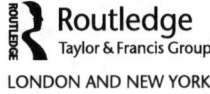

LONDON AND NEW YORK

First published 2023
by Routledge
4 Park Square, Milton Park, Abingdon, Oxon OX14 4RN

and by Routledge
605 Third Avenue, New York, NY 10158

Routledge is an imprint of the Taylor & Francis Group, an informa business

© 2023 Janice Baker

The right of Janice Baker to be identified as author of this work has been asserted in accordance with sections 77 and 78 of the Copyright, Designs and Patents Act 1988.

All rights reserved. No part of this book may be reprinted or reproduced or utilised in any form or by any electronic, mechanical, or other means, now known or hereafter invented, including photocopying and recording, or in any information storage or retrieval system, without permission in writing from the publishers.

Trademark notice: Product or corporate names may be trademarks or registered trademarks, and are used only for identification and explanation without intent to infringe.

British Library Cataloguing-in-Publication Data
A catalogue record for this book is available from the British Library

ISBN: 978-0-367-74182-2 (hbk)
ISBN: 978-0-367-74205-8 (pbk)
ISBN: 978-0-367-74194-5 (ebk)

DOI: 10.4324/9780367741945

Typeset in Times New Roman
by Newgen Publishing UK

Contents

	Acknowledgements	vi
	Introduction: Museums and transformation	1
1	Disappearing soils: Toward a pithier pedagogy	20
2	Sinking and melting: Glossing the climate problem	35
3	Repurposing the inclusive museum	51
4	Museums, climate fiction and the anthropocene	64
5	White geology and displays of material power	78
6	Coal and fossil capital	88
7	Oil utopias and petro-invisibility	100
8	Museums inside the earth	112
9	Gold on show: The toxic glamour of the yellow rock	124
	After neutrality: The relevant museum	133
	References	143
	Index	157

Acknowledgements

I acknowledge the Wadjuk Noongar People, traditional custodians of the land on which this work is written, and pay my respects to elders past, present and emerging.

I thank the anonymous reviewers of this book, which has benefitted from your advice and suggestions.

Introduction
Museums and transformation

Museums are uniquely positioned to provide meaningful governance on climate change by how they frame relations with collections, objects and events. Museums transform how we feel and think on issues and can boldly confront the climate emergency. A way to do this is to differently communicate our kinship with soils, minerals and fossil fuels. Alternative approaches to these actants as resources for production and consumption are needed and this suggests a new type of inclusion that shifts our understanding of what it means to be human. Museums can no longer claim the prerogative to remain neutral; the times are too fragile and the climate emergency is here now.

In a time of disunity, anxiety and misguided nostalgia, museums are uniquely positioned to shift the western imaginary to acknowledge the agency of material actants as part of life. In this, the science that understands physical systems is vital but science by itself is only one part of the story. Climate is statistical data collected and modelled over time by scientists, but how this data are absorbed is causing conflict and confusion socially, economically and psychologically. The philosophy underscoring how we think about life has to change:

> It is only possible to think of climate change in the meteorological sense – with humans now bound to volatile ecologies that they are at once harming and ignoring – if some adjustment is made to the ways in which we think about the relations among time, space and species.
> (Colebrook 2014, 10)

That climate modelling and extreme weather events require institutions to 'act now' through changes in policy and practice is no longer news, as confirmed by Reports of the three Working Groups of the Intergovernmental Panel on Climate Change (see www.ipcc.ch/assessment-report/ar6). As museums have a legacy of authority in the research and presentation of both science and culture, it is significant that they unequivocally act now in presenting more just, equal and open relations with the earth. Climate security researcher Elizabeth

DOI: 10.4324/9780367741945-1

Boulton advises, 'It is highly significant which stance on climate change (act now, later, slowly, or never) becomes the dominant narrative' (Boulton 2016, 774).

This book was written in tandem with extreme climate events and a global pandemic. From Western Australia where I live, it seemed the 'forever' bushfires of 2019/20 in the east of Australia would never end. The fires were 'like nothing previously recorded anywhere on Earth' (Flannery 2020, 57). Billions of animals and a sixth of the vast continent burnt. The immensity of the tragedy is difficult to comprehend. Fires were followed by the arrival of the corona-19 virus and the closure of the state border for 700 days. The end of 2022 COVID isolation in Australia was accompanied by multiple massive destructive flooding events on the east coast.

These cumulative events, losses and vulnerabilities informed my musing on the significance of material thinking during these difficult times, and firmed the awareness that the excessive consumption culture in the global north must end. Through a focus on the 'inhuman', I seek a role for museums in a fundamental transformation in thinking. In this my ideas are often speculative and I thank the reader in advance for their forbearance as I try to match philosophy with politics and practice.

The book has been both limited and sharpened by disruption. Unable to travel – which in relative terms was not a hardship on par with most people's experience – my ideas assembled through recollection of encounters in museums, galleries and heritage sites, augmented by museum and critical heritage case studies and scholarship, evaluation of museum websites and insights offered by contemporary art. What has become clear from this endeavour is that there are museum visitors, workers and supporters, like myself, concerned that museums take an active role in meaningful climate justice and governance. Others, echoing public disinterest or apathy, prefer that nothing changes. But the latter voice is diminishing. Amidst the spectrum of views and values, those bold museums that respond courageously to the challenge to act now will carry the institutional relevance of museums into the future.

There are diverse ways to tell the story of institutional climate inaction and denial. Armed with insight into the deceptions used by the tobacco industry (exposed in their 2010 book *Merchants of Doubt*), Naomi Oreskes and Erik Conway (2014) combine current climate science with an imaginary future should no global action be taken. Their speculative nonfiction is located in the near present following the crash of western civilisation, and takes a clear stance on the impact of failing to confront the fossil fuel industries by reducing emissions.

> The dilemma addressed is how we – the children of the Enlightenment – failed to act on robust information about climate change and knowledge of the damaging events that were about to unfold …
> (Oreskes and Conway 2014, ix)

A developing genre of near-future dystopian climate fiction (cli-fi), such as Oreskes and Conway's, confronts climate breakdown and the implications of living in the Anthropocene. Adopting a dystopian stance is only one take on climate; at the other end of the spectrum are techno-utopian discourses that assume that ingenuity to innovate will not only 'fix' the climate, but in the process augment a more advanced humanity. Such trans-humanist discourse positions humans at the centre of everything and perpetuates our exceptionalism by segregating organic actants (us) from the inorganic and lithic (it). The artificial segregation of *homo sapiens* from the matter of the earth, from geological actants, has enabled the tremendous exploitation of soils, minerals and fossil carbon since the industrial revolution. The segregation of the organic and inorganic is a construct that must be rethought. This requires a more nuanced attention to the meaning of inclusion that enables deeper thinking on geological-human entanglements.

Material thinking is an ethical rather than hubristic adaptation to climate change. On the dystopia-utopia climate spectrum, however, museums are inclined toward techno-utopian fixes that maintain the status quo. In effect, this supports the view that economic growth is the way out of the climate quandary and for museums equates to remaining neutral. Given predicted disruptions accompanying weather events, a museum that adopts a 'wait-and-see' stance will lose its relevance to citizens and communities, and find they are on the wrong side of history.

There are astute evaluations of climate-related exhibits and the role of museums in climate change governance and I engage with this body of research in the chapters ahead. One of these projects on climate and museum futures, collated by Fiona Cameron and Brett Neilson (2017), concludes that museums 'have a role to play in communicating the full complexities of science and providing new perspectives on climate change as a complex scientific, cultural, economic, and social issue' (8). The researchers confirm that despite the difficulties confronting museums and science centres, they 'are important and powerful venues in climate change governance and in coordinating a global response' (7).

One of my interests is to understand how museums are engaging or can engage with some of the more challenging insights into material thinking. I'm interested in understanding the museum's legacy through what Kathryn Yusoff (2018) calls a White Geology, and how this legacy aligns with fossil capital (Malm 2016). There are many exhibits dealing with the theme of the Anthropocene that do not deeply probe the material conditions these and other critical concepts enable. It matters greatly how museums communicate the degradation and disappearance of the earth's carbon storing soils, the environmental impact of extracting and processing minerals and the burning of carboniferous fossils.

These historical and current events are enabled and largely controlled by global multinational companies whose massive profits are difficult to extricate from shadow governments and dependent state actors. This is the planet's difficult heritage. A heritage that makes degrowth a contested and complex action, while clearly, change has to happen:

> The task is to pull back from the precipices that growth-oriented production ignores or does not understand, to do what we can to undo the damage already done, and to reconstruct our economic relations with one another and with the non-human world.
>
> (Mann 2022, 27)

In musing on geo-inclusion – an inclusion of material agency – a geo-philia (love of rocks) emerges that acknowledges life as fundamentally an assemblage of elements; this awareness can infiltrate with greater nuance into thought. It is known that life is recycled carbon, but as an abstract understanding rather than awareness that consciousness is embodied. I am drawn to contemporary art and literature that observes life as carbon being that responds to thresholds between things usually considered separate and that acknowledges the dangerous hubris of the human species.

Identifying the problem

Responses to climate (past, present and imagined) are approached in the chapters ahead through the dilemma of the inclusive museum. I wonder whether the ubiquitous adoption of inclusion is adequate to the climate challenge. Can the focus on the cultural identity of a subject that defines practices of 'inclusion', 'access' and 'participation' attend to structural changes that acknowledge our inhuman grounding in time and space? How might museums recalibrate their own self-interest as a starting point for a larger move toward the agency of a collective and more-than-human reality?

These questions direct toward philosophical depth in museum theory and practice. Museums are capable of critiquing posthuman, posthumous and inhuman futures through repurposing their collections and practices as the object of enquiry. A shift that does not herald a new museum as such, but perhaps observes hidden, in-waiting, emergent museums. There are already non-museums acting as counter institutions to the traditional idea of the museum. Colin Sterling cites, as examples, the Museum of Nonhumanity (2016) and the Museum of Capitalism (2017).

> The *museum that is not a museum* acknowledges the complex history of such institutions as a first step toward marking out a space of critique *inside* this tradition. The very nomenclature of "the museum" is vital to this work, immediately invoking a set of spatial, discursive and aesthetic conventions against which a counter-proposal may be registered.
>
> (Sterling 2020, 194)

Communicating that soils, minerals and fossils are not mute objects does not conform with western knowledge schemas that split nature from culture so effectively that the threshold between them cannot be breached. The split is so effective that the social and biological human is defined by not being nature. It is obvious that this separation is an artifice, yet it remains dominant. But it is always possible to breach assumptions that have become common sense. The idea of nature as somewhere out there and physically apart from us can be retold in ways that are commensurate with the entanglement of life with the impacts of the meteorological: rising seas, melting ice and permafrost, drought and fires, floods and storms, loss of biodiversity, species extinction, health decline, eco-depression …

Museums that are relevant are those that admit a revised common knowledge based on awareness that for all its unique brilliance, western reason is devised of epistemologies and ontological stances that are not responding well to the climate emergency. Museums must become sage influencers of difference. Institutional sagacity has been muted by a legacy of participation in the instrumentalisation of nature, and this is why material 'turns', and related approaches linking the sciences and humanities are useful. Material and relational turns expose the constructed character of oppositional thinking that separates human life from more-than-humans (oil, weather, fungus …).

This does not go unacknowledged. Fiona Cameron's idea of a 'liquid museum' embraces ecologising logic as a method to break the stranglehold of modern human-centred views of climate change and the environment (2017a, 29). Bringing cultural theory into the fold, she adopts liquid thinking as a means to 'disrupt the stolid and solid imaginary of the modern museum and its hard, disciplinary, authoritative powers and reformatory agenda' (2017a, 26).

Another approach is curator Tom Jeffery's adaptation of critical realist philosophy to 'disrupt the persistence of museological human-nature dualism' (2021, 48). In his analysis of the neoliberal commodification of museum and heritage practices in South Africa, Jeffery conceives a social-ecological activist position for museum work, with the capacity to directly challenge the dualisms that exclude the museum from real ecological work. While, in the field of heritage, Rodney Harrison (2015), who acknowledges heritage as multiple overlapping ontological fields, presents a 'new heritage' that 'has nothing to do with the past at all' but is a form of futurology that folds the human and non-human to offer new entanglements of care (35). These few examples are reflective of directions for critical museum and heritage theory and practice that communicate a revised common knowledge of museums and climate relations.

Elizabeth Boulton (2016) affirms that 'for many theorists, climate framing sits at the highest philosophical level, of understanding and representing a new reality and enabling people to "think" and "exist" differently' (775). On this, Boulton evaluates the efficacy of Timothy Morton's oft-cited concept of the climate and its effect as a hyperobject. For Morton,

> A hyperobject could be a black hole. A hyperobject could be the Lago Agrio oil field in Ecuador, or the Florida Everglades. A hyperobject could be the biosphere, or the Solar System. A hyperobject could be the sum total of all the nuclear materials on Earth; or just the plutonium, or the uranium. A hyperobject could be the very long-lasting product of direct human manufacture, such as Styrofoam or plastic bags, or the sum of all the whirring machinery of capitalism.
>
> (Morton 2013, 1)

As an ecocritical-influencer Morton addresses the difficulty of communicating climate by conceiving global warming as a vast object that exceeds apprehension yet that everyone senses. Morton's hyperobjects have been reviewed as scary game changers that carry a touch of the sublime: 'Like Bob Dylan's Mr. Jones, we sense something is happening, but we don't know what it is' (Muecke 2014). That climate is a hyperobject that cannot be grasped in its entirety is what enables climate deniers and the fossil fuel lobby to claim the voice of common sense.

Here is where museums can take the reins and communicate what is meant by a hyperobject, not as a way to grasp and order the chaos, but to respond to community feelings of helplessness, to apprehend why most people feel unable to act. Museums can convey that they too are weathering the climate of change.

Why include the inorganic?

Nuancing the meaning of inclusion is a method to better value that any agency humans have to act is joined to the inorganic. Many museums might presume they defer to the material realm already, and that their researchers, collections and pedagogic practices make them experts in such communication. Yet museum displays continue to position humans as the main game, reflecting an anthropocentrism that at a fundamental level represents neutrality when it comes to new thinking as an action and adaptation to climate change.

The focus on inclusion dwells particularly on the habit of excluding inorganic agency, which has led to horrifying events of forever fires, sinking islands and disappearing soil. Admitting that soils actually define life involves departing from certainty about boundaries and thresholds between things. It requires that museums reflect on the role they perform in defining the taxonomic certainties that categorise nature and enable its commodification.

A careful evaluation of the impact of practices that may appear harmless is necessary. In January 2020, the Natural History Museum in London declared a planetary emergency (nhm.ac.uk) embedding its response to this emergency as the NHM Strategy to 2031. Surely a good news story. Yet the strategic plan does not perform as an altered response to nature. The plan talks about connections with nature, and inclusion, but in so doing carries the conventional separation of humans from the material world. 'Our vision', it says, 'is of a future where both people and the planet thrive'. At a subtle level what the rhetoric communicates is that people and the planet are separate. And it does not convey that the planet does not give a toss about people.

Cameron (2017b) has surveyed merchandising at the NHM and observes that its use of branding slogans such as 'We are on Nature's side', 'Nature with a big N' and 'Nature's Treasure-house', while seemingly harmless,

> suggest that humans are there to protect Nature; are custodians of nature in a way that displays hubris and claims to patriarchal continuity; that museums as modernist institutions give themselves the authority to speak for the nonhuman; and that museums act as storehouses for all the kingdoms of Nature.
>
> (51)

The inclusion dilemma has to be acknowledged in order to shift the playing field to confirm that lithic entities and physical systems share the planet and biosphere with that very recent arrival, *homo sapiens*. Not the other way around. What this opens up is a more nuanced enquiry such as Tim Ingold's desire that 'If humans can study the earth, why can't the earth study humans?' (2021, 121). Nuanced approaches to inclusion – with alterity as part of human thinking and not its limit – is not absent in the western imaginary. There are stories of wondrous soils and stones in ancient and medieval texts, creation mythology, curiosity cabinets and Wunderkammer, science fiction and contemporary art.

Exploring alterity does not advance museums as hosts for material thinking as a pseudo-scientific, eco-romantic fantasy. On the contrary, it is a goal of the book to resist the naïve optimism found in types of relational thinking that extol an unbounded connectivity between people and nature. Naïve connections with the earth are discernible everywhere – in the happiness and wellness industries (Davies, 2015), in techno-utopianism, in neoliberal managerialism and innovation rhetoric of universities, in the greenwashing of commodity culture, in appropriation of First Nations motifs and in social media and Twitter triteness. Unbridled optimism in the claim that everything is connected to everything else does not communicate a structure of understanding and sensibility commensurate with climate disruption; rather such eco emotion is a hyperbolic form of conspicuous compassion.

A shift in rhetoric and practice toward institutional reflexivity – instead of accommodating the neoliberal museum of late capitalism – involves acknowledging the value of critical thinking. What used to be known as the new museology can repurpose its agenda, and acknowledge climate, its disruptions and adaptations as the central justice and equity issue for museological reform. The inclusive museum can be emancipated from neoliberal attachments to hyper individualism as a desirable identification for the human.

The Janesian call

In 2009 museum activist Robert Janes cautioned that museums are dangerously out of touch in failing to address issues confronting 'a troubled world'. In 2020, with the rising awareness of global warming and environmental crises, Janes continued calling for action, frustrated at flawed governance and leadership in museums, and arguing for courageous and creative action. 'For reasons that remain unclear, the international museum community has not yet had a forthright conversation about the climate crisis, including the role and responsibilities of museums, which would lead to concerted action' (Janes 2020, 587).

Neutrality is a stance rejected by activists working across the museum sector: 'For museums to matter in a time of climate crisis, they must first reject the claim to political neutrality that structures and limits their transformative social power' (Lyons and Bosworth 2019, 174). The notion that museum professionals remain neutral and not reflect bias in relation to their expertise belongs to a defunct museum regime. Museums can usefully assert an unambiguous criticality, most obviously in response to the burning of carbon and the power wielded by fossil fuel industries. As Janes and Richard Sandell (2019) write, 'it is time for the global museum community to speak as clearly and forcefully as its privileged position in society demands from it' (18). It is a position many museum workers share. Tom Jeffery's survey of South African museum professionals identifies 'a tension between respondents' individual, progressive social-ecological concerns and the neoliberal ideology of the institutions for which they work' (2021, 59).

An absence of critical depth and comparative cross-disciplinary perspectives within the discipline hinders confidence to take action, 'As a field, critical museology still remains an extraordinarily undeveloped subject of study' (Shelton 2013, 481). And for Cameron (2017a), 'much work is yet to be done to make museum theory and practice more relevant for the present-day, and for promoting new styles of thought and action' (23). Addressing the critical lacuna, cross-disciplinary approaches highlight alternative world-making to museums' humancentric orientations. Critical perspectives on de-colonising and re-territorialising sites and objects validate a reflexive focus becoming habitual in museums.

The limitation of assuming that the role of museums is to uphold the civilising tenets of modern and neoliberal humanism is apparent if we look closely at practices of collection, categorisation and exhibition. In the case of the lithic, interpretation of elements for their human use prevents a more sustainable thinking that 'sees' the agency of geological existents and notices the effects of extraction economies. In his book *Stone* (2015) medievalist Jeffrey Cohen communicates the type of agency that stone enables:

> Stones are rich in worlds not ours, while we are poor in their duration. We therefore have a terrible problem communicating with each other.

Geophilia marks an enduring inclination to lithic alliance, an embrace of coextensiveness, despite or perhaps because of the cognitive reeling stone can trigger.

(Cohen 2015, 249)

The unsettling poignancy of encounters that return agency to the lithic upsets the instrumental rationalism that props up the global industries that profit massively from extracting and trading minerals and fossil fuels. I seek interactions with rocks, soils and fossil fuels to investigate the anthropocentricity that disables our ability to imagine the elemental liveness that enfolds life. The idea that our clever minds are separate from the elements of the earth, from the aliveness of stones and sediments, is a wrongheadedness that supports growth economies that it is foolish to uncritically perpetuate.

Rock sites

My musing on stony entanglement is impassioned by the destruction of First Nations Peoples rock sites by mining companies in north Western Australia. Most recently, in 2020, Rio Tinto dynamited the Juukan Gorge shelters of the PKKP (Puutu Kunti Kurrama and Pinikura People) in the Pilbara region of the state. The cave system in the Pilbara that the mining company destroyed connected the PKKP through ancestors to all things in Country and beyond. A shocking aspect of this destruction is that mining giants continue to escape punishment commensurate with their crimes.

The shelter destruction marks a new measure for action by responsible museums, certainly in Australia. A museum cannot be neutral toward the state power that supports Big Mining. Constant vigilance is required. Sparked by the Rio Tinto disaster, in 2021 the rights of Indigenous people over use of their Country for mining was put to the test. It was ultimately neutralised with the final say over cultural heritage given to the Minister for Aboriginal Affairs. So while a new Aboriginal Cultural Heritage Council and Cultural Heritage Bill has been established, these laws perpetuate relations between colonised and coloniser, a relation based on economic growth and exploitation of the land (Trigger and Goerling 2021).

Geographer Tod Jones (2015) observes that the systematic discrimination against Aboriginal heritage in Western Australia comes not from an identifiable 'racist administrator' but from the 'evolution of the institutions, rules and conventions that make up cultural heritage management'. Part of this accumulated institutional failure is the philosophical inability to admit the entanglement of human and more-than-human agencies. There is a failure to acknowledge and confront the impact of 'organic chauvinism' – a particularly western arrogance that overrides all forms of inorganic existence. Such careless intention is bolstered by a rhetoric of inclusion that excludes the lithic.

For First Nations the destruction of rock sites, outcrops and landscapes *is* murder. In relationships with Country, Aboriginal people don't 'abide by any fundamental difference between Life and Nonlife' (Povinelli 2016, 102). The same is true of water on Sioux land threatened by the Dakota Access Pipeline, an act for the Dakota and Lakota, 'no less genocidal that the eradication of the people themselves' (Deem 2019, 126). These ontological relations with the earth necessitate a psychological and emotional awareness of acts and events that cannot justify mining extraction and supply activities.

It is increasingly apparent that extracting ores and carboniferous fossils, and shipping or piping these earth agents to global markets, perpetuates a business model and way of life that is no longer sustainable. The 2022 Russian invasion of Ukraine has exposed Europe's reliance on Russian energy sources arising from slowness, obfuscation and disunity in the transition to adequate and sustainable energy sources.

In stark contrast to unfettered industrial growth, First Nations Peoples engagement with the same earth is mutually agential, a relationship developed across generations that does not separate the cultural, social and psychological. The responsibility of Indigenous communities to the flow of materials is ongoing and cannot be cancelled or voided without social dislocation that ripples through Country. Forms in the landscape require attention. For Indigenous people in northern Australia, the attention will manifest 'when something discloses itself as comment on the coordination, orientation, and obligation of local existents and makes a demand on persons to actively and properly respond' (Povinelli 2016, 58). There is a proper way of doing things.

In communicating that the weather is disrupting ways of life because of human exploitation of materials, there is much to be learned about water, fire, air and earth from First Nations commitment and knowledge to Country, a responsibility ontologically missing in the western mindset. Interdisciplinary anthropologist Deborah Bird Rose observes that the state of the country,

> ... offers concrete evidence of the responsibility which the owners have been exercising. Responsibility is grave: there is no hiding in a conscious universe ... the exercise of will in a situation where the choice to deny moral action is to turn one's back on the cosmos and ultimately on one's self.
>
> (cited in Pascoe 2014, 180)

Writing of broken landscapes and the unusual behaviour of whales in Northern Australia, Tyson Yunkaporta (2021) conveys the importance of knowing stories that speak the law that is the land. The purpose of knowledge is to tell the right stories about the sentient landscape.

> Proper story is a living landscape model that allows you to make accurate predictions, shows you the limits and obligations of your relationship with the land. It is a collective, an aggregate of the knowledge of many people who speak for different aspects and diverse bioregions.

Indigenous knowledge is distributed and non-centralised and aligns, 'with the complex ecosystems we have inhabited over hundreds of millennia' (Yunkaporta 2021).

I am cognisant in respecting First Nations knowledge to not generalise what are complicated and diverse cultural experiences. I'm aware of writing about other cultures and peoples through the lens of my western mindset. However, the times are urgent and I intend these examples of connections to Country to support a methodology for thinking materially in ways thwarted by western value systems.

The physical, emotional and psychological devastation created by mining and its disturbance of traditional relations between Indigenous communities, entities and ancestors are neither new nor unusual. It continues a form of geo-bio-cide that is the settler colonial project. Murujuga (Burrup Peninsula) along with much of the Pilbara region in north Western Australia is both sacred and a site of massive mining operations, particularly iron ore and gas (Baker 2022). The region as an economic resource (it has been called the engine room of the Australian nation, and a Chinese quarry) exemplifies the extraordinary global impact of multinational corporations and gives resonance to naming the era Capitalocene.

From her experience of living with the Belup peoples in northern Australia, anthropologist Elizabeth Povinelli articulates the ontological incommensurability of Indigenous knowledge and western versions of this knowledge. She has investigated mining destruction across the north, such as OM Manganese Ltd. at its Bootu Creek mine of the sacred site known as Two Women Sitting Down. In mining terms this site is a manganese outcrop; however for local people the manganese is the blood of ancestors. The company was successfully prosecuted for its intentional destruction of Two Women Sitting Down, but this did not lead to any alteration of mining practices but rather to an attack of such lawsuits (Povinelli 2016, 32).

Povinelli confronts the damage caused by activities supported by late liberal ideology, and articulates the need for geontological awareness where the life of rock and other Indigenous knowing are not diminished and negated by the usual fallback to the Life and Nonlife divide. She argues for multidisciplinary perspectives, a new interdisciplinary literacy, a gathered wisdom to tackle the Anthropocene. Otherwise, she notes, the geological age of humans 'will be the last age of humans and the first stage of becoming Mars, a planet once awash in life, but now a dead orb in the night sky' (36).

Museums are sites of gathered knowledge, but what happens when what is known is no longer wise enough, when there is not proper thinking? An admission needs to occur. And the hubris dropped.

Art history

Since the 1980s, clamour for more inclusive and accessible museums has drawn encounters with difference away from the other-than-human. Difference and otherness have been framed around diverse and marginalised individual and group identity in ways that interpolate 'the other' into the system that needs transforming. This is difficult territory that needs careful critical consideration as there is nothing to be gained by diminishing the equality sought by supporters of the marginalised. The dilemma arises when cultural diversity is politicised to support neoliberal management of difference.

The now mainstream management of identity politics is not unrelated to a lack of environmental criticality across fields of inquiry, including in art history. A lack acknowledged by environmental and art historians Andrea Gaynor and Ian McLean (2005) in a study on the limits of art history:

> The reason for the art world neglect of ecology goes deeper than an ignorance of, or disinterest in, the science of ecology. The popularity of the issues of race, class and gender reveal how anthropocentric the art world is. As a discipline and practice, art history descends from humanism, and since its formulation in the nineteenth century, it has mainly addressed questions of nationalism and identity that, in western art at least, presume a theory of the human subject that is no longer sustainable, either environmentally or scientifically.
>
> (8)

Since 2005, when this was written, there have been critical moves to address the complex relations between ecological thinking and art history. The influence of interdisciplinary texts, such as Heather Davis and Etienne Turpin's *Art in the Anthropocene* (2015), Anna Tsing et al.'s *Arts of Living on a Damaged Planet* (2017) and Mitman et al.'s *Future Remains* (2018) reflect creative engagements that step into territories beyond humanism and cultural identity. However, the status quo remains dominant, framing climate as a topic of interest rather than the central concern for life on the planet. For art historian Andrew Patrizio (2019), 'Ecological concerns have traditionally been so little addressed in the art historical canon that even their absence has rarely been noted' (29). A recent collection of essays surveying intersections between digital humanities and art history (Brown 2020) would appear to exemplify this tendency. The book's introduction outlines legacies of gender and racial prejudice relevant to the themes of the book (1) but does not mention the digital humanities intersecting with climate crises. A way of interpreting such editorial absence might be a text's assumption of a technological utopia. Techno-utopianism appears in many guises averting or obscuring that it sits within the problem of human exceptionalism.

A too narrow vision of what a museum can be limits encounters with alterity. Here we can bring to the museum modes of understanding opened by contemporary art, for example, Ilana Halperin's explorations of a corporeal mineralogy that engages the boundaries between the body and geological (Halperin 2015, 79). Or Tomás Saraceno's flying solar sculptures built of plastic bags that provide the experience of becoming aerosolar. These flying museums have a sensitivity to the air that is 'an invitation to think of new ways to move and sense the circulation of energy' (Saraceno et al., 2015, 59).

Art historian T.J. Demos (2016) sees climate change as a political crisis that requires the will to change. Contemporary art shows that the decolonising of nature is at the forefront of the transition. We see here a different mode of awareness to that of modern art, which used shock as a form of defamiliarisation to force a willingness to change. As Davis and Turpin (2015) acknowledge, 'Beyond the modernist valorization of the principle of shock in art, our current climate demands a different kind of aesthetic and sensorial attention' (11).

How art and museums are relevant (or not) in a post-shock world is increasingly interrogated in the field as conveyed by an activist member of Zone à Défedre, a group effective for many years in preventing the French state building an airport north of Nantes.

> Why design a dance piece about food riots when your skills as choreographer could help crowds of rebels move through the streets to avoid the police? Why make an installation about refugees being stuck at the border when you could design tools to cut through the fences? Why shoot a film about the dictatorship of finance when you could be inventing new ways of moneyless exchange? Why make a performance about the silence left where there were once songbirds when you could be creating an ingenious way of sabotaging the pesticide factories that annihilate them? Why continue to feed the zoo of the artworld when your creativity could be stoking the rebellion?
>
> (Jordan 2021, 390)

William Fox (2017) describes recent art as an intervention, observing a new consonance between the artistic and scientific. 'There has been a movement of artists from picturing nature into imagining the human footprint upon it, and then to intervening in the systems of the world' (197). Demos' perspective is along the same lines:

> … we cannot address climate justice adequately without also targeting the corruption of democratic practice by corporate lobbying, or the underfunding and failure of public transportation systems, or Indigenous rights violation by industrial extractivism, or police violence and the militarisation of borders. For these areas all link up in one way or another as interconnected strands of political ecology.
>
> (2016, 12)

Apathy

In the twenty-first century gross inequity is openly acknowledged. The richest 62 individuals possess as much wealth as the 3.5 billion poorest (Oxfam 2016).

14 *Introduction*

The facts are alarming, 'the products of 100 companies are responsible for 71% of the world's greenhouse gas emissions' (Lyons and Bosworth 2019, 178). Facts and data should blast the ventures of late capitalism and neoliberal economics apart, yet belief in free markets, growth as progress and individualism continue to dominate and perpetuate local and global carbon economies. The duplicity of 'free' market economics is well exposed, exemplified by massive government subsidies for fossil fuel capital. Climate might be a hyperobject but (despite being unpacked by Marxist and neo-Marxist theory) so too is capital. 'The one vector changing the composition of the atmosphere and the earth's crust, pushing socioecological systems over the brink, fantasizing about colonizing Mars, all the while mostly escaping critical scrutiny and democratic transparency, is plainly capital' (Saldanha 2017, 231).

Although many are aware that much is awry, clarity around climate change and ecological catastrophe is thwarted by misinformation. Lack of clarity excuses voters to support a politics that offers the lifestyles required by fossil capitalism. Demos (2016) argues it does not need to be this way and that 'there are plenty of solutions for sustainable living today, which, if implemented globally, could protect biodiversity and define a more equitable and inclusive socioeconomic order' (12).

Critical thinkers are confronted by the power that denialism and apathy have to organise politics. For some the scenario tips into the mental health condition of solastalgia, the psychological distress of having no control over climate and environmental events (Albrecht 2015). Andreas Malm (2016), the climate scholar who coined the term fossil capital, says, 'It is proven beyond all doubt that global warming does not have natural causes. Solar radiation, volcanic outgassing, endogenous variations in the carbon cycle, and other similar suspects have been decisively cleared of responsibility for the rise in temperatures' (21). Yet we find climate doubters and deniers turning to notions such as solar radiation rather than confronting the need to change the thinking that begets our separation from the natural world.

It is not enough for museums to document in an objective and non-committed way the findings of climate science. There is 97% agreement so waiting for scientific consensus is not only pointless but also counterproductive: 'politics not science, must take centre stage' in the museum (Hulme 2017, 11). For geographer Mike Hulme, 'The myth of consensus ... is perhaps the biggest problem facing climate change politics today. Assuming political consensus as a horizon marginalizes those who dissent and undermines the role of disagreement of politics' (12). It is also the case that climate change has so many meanings that it has created 'a political log jam of gigantic proportions' (Hulme 2009, 333).

The impasse of either the political right or left to adequately deal with climate change and the function of globalisation in the new climate regime is framed by philosopher and sociologist Bruno Latour with the notion of the 'Terrestrial' as an alternative way of thinking with the material world. The Terrestrial signifies a return/turn to the Earth, a way of thinking that is neither local nor global. The Terrestrial looks inward yet is not aligned to individualism and identity. Latour views this material turn as a way of worlding, which is Donna Haraway's term for refusing borders or identities. The need is urgent to support the radical Terrestrial, Latour (2017) argues, 'before the militants of the extreme Modern have totally devastated the stage' (56). The once-modern museum can become a Terrestrial zone, a vanguard of attention to the more-than-human as pivotal in all engagements with collections.

Latour is one of many theorists for whom the notion of progress is disingenuous. This said, critics observe that Latour's engagement lacks the ontological awareness that already exists with First Nations historical and cultural experience. Zoe Todd is 'critical of the sort of post anthropocentric "cosmopolitics" Bruno Latour proposes and his failure to acknowledge Indigenous epistemological precedents which have long forwarded a radical climatological vision' (Deem 2019, 125).

Nevertheless, new terms are helpful to set in motion alternative thoughts and ideas, to provide alternatives to modernising terms. With climate change many ideas are under the microscope, including 'world', 'nature', 'global', 'Anthropocene' and 'humanism'. The earthling is becoming Terrestrial in ways that question the relevance of dominant western concepts and metaphors. Global warming requires concepts to challenge previous certainties, terms to mark out new images of thought. Acknowledging new terms and inventing neologisms, even while these may initially appear strange, communicates that there are alternative concepts being explored to counter or reject dominant humancentric discourses.

Humanism is itself a tricky keyword that conjures images of culture and civilisation, attached to ideas of self-development and perfection (Raymond Williams 1976, 150). Keywords are perhaps the linguistic equivalent or marker of hyperobjects in being too vast and vague to contain. Philosopher Rosi Braidotti (2013) explores how the Humanist model is a standard for both individuals and for culture, a civilisational model that has 'shaped a certain idea of Europe as coinciding with the universalising powers of self reflexive reason' (13). The perfect ideal of 'Man' as a representation has come to personify a boundless capacity of humans to pursue their individual perfectibility (13). Within this frame of action, difference comes to mean that which is beyond the bounds of reasonable, and this includes the material world in all its forms and manifestations. 'Subjectivity is equated with consciousness, universal rationality, and self-regulating ethical behaviour, whereas Otherness is defined as its negative and specular counterpart' (15). To escape the hubris of hyper humanism, the images we construct of the human, the human we recognise, must transform.

Action

Museums are vested with authority to take a leadership role in new ontological awareness. While leadership in large museums has often been self-serving, there are examples of involvement in advocacy supporting frontline activists. In 2014 the activist art collective Not An Alternative founded a nonprofit programme called The Natural History Museum, a programme that 'works at the interaction of art, activism, cultural organizing and critical theory' (The Natural History Museum). NHM's mission cites the museum will,

> ... inquire into what we see, how we see, and what remains excluded from our seeing. It invites visitors to take the perspective of museum anthropologists attuned to the social and political forces inseparable from the natural world.
>
> (ibid)

In 2016, NHM organised an open public letter to President Obama and relevant governmental agencies to protest the destruction of sacred sites of First Nations in the construction of the final section of the Dakota Access Pipeline. The 1,886 km underground oil pipeline begins in the Bakken oil fields in Northwest North Dakota, travels through South Dakota and Iowa, ending in Illinois. The destruction of First Nations land by Energy Transfer Partners is reminiscent of the ubiquity of destruction that attends mining activity, like Rio Tinto's dynamiting of Aboriginal sites in the Pilbara.

Considered 'an amazing act of solidarity' by the Sacred Stone Camp blockading the oil pipeline, the open letter included 'fifty executive directors of museums and institutions of archaeology or anthology (including the Smithsonian Institutions, Washington, D.C, the Field Museum, and the AMNH)' (Lyons and Bosworth 2019, 179). Signing the letter was experienced as a cultural shift for museum leaders, many recognising 'the urgency of leveraging their influence and expertise to support those working hardest to fight the corporations most responsible for anthropocentric climate change' (ibid., 181).

Change is happening, and museums startled out of complacency. Museums can be profoundly insightful when they contextualise the unpalatable meanings of their collections. An early, frequently referenced example is the exhibit curated and installed by Fred Wilson *Mining the Museum* (1993) that revealed the Maryland Historical Society's historical relation to slavery through rethinking the metal ware in its collection, notably iron slave shackles and silverware made by slaves (Karp and Wilson 1996). Echoing this, in 2016, The Natural History Museum project *Mining the HMNS* interrogated the symbiotic relationship between the Houston Museum of Natural Science and its corporate sponsors. The project investigated Houston's fossil fuel ecosystem including 'co-hosted monthly Toxic Tours of East Houston's petrochemical plants and refineries' (Lyons and Bosworth 2019, 181).

In his project *Uncanny Sensing, Remote Valleys*, as part of the exhibition *Placing the Golden Spike: Landscapes of the Anthropocene* (Hannah and Krajewski 2015), artist and researcher Steve Rowell investigates and documents toxin- and chemical-affected environments around Houston. He observes the reliance of humans on non-human technology, and ponders the use of technology to mitigate climate change. His image of an exhibit in Houston Museum of Natural Science shows colourful neon tubes intended to 'educate future oil workers by illustrating the various working complexities of a petrochemical refinery'. He points out that sponsorship of the display is a 'who's who of the industry and include: Saudi Aramco, Halliburton, Shell, ChevronTexaco, ExxonMobil, American Petroleum Institute, Enron (now dissolved), ConocoPhillips, and British Petroleum' (cited in Hannah and Krajewski 2015, 94).

It is artists who first exposed the business of museum relations with the multinationals and corporate giants who manage and perpetuate fossil capital. While contemporary artists move outside the museum to intervene at places of climate disruption, the museum must turn its newly inclusive eye toward the changing geo-bio subject of climate change politics.

Museums' commitment to inclusion can productively shift in focus – a lithic turn that perhaps only the museum is given the trust to do – toward earth's elements as good to think with. This entails revealing whenever logic segregates nonlife and life in modern thinking enabling fossil fuel carbon-based ideology, rhetoric and lifestyles. In thinking through this task, I don't attempt to create a new philosophical argument. The works of many astute thinkers already achieve this. The arguments are extant in philosophies of difference, contemporary art, new materialism, object-oriented-philosophy, speculative realism, posthumous theory and other critical and creative engagements with materials that cross boundaries between art and science to push thought beyond the limits of humanism.

The destruction caused by Big Mining is heartbreaking, everywhere. Rio Tinto destroyed Country in use by the PKKP peoples for at least 46,000 years. The Black Mesa, the largest coal deposit in the United States, was strip mined for 40 years devastating Navajo (Dine) environments and communities. Water and coal are agents within Navaho cosmology. The waters are the mother's blood and the coal is her liver. But indigenous geo-bio social anatomy is utterly illegible to Peabody Energy and to settler colonialism more broadly (Haraway 2016, 96). In southeastern Montana, Naomi Klein (2015) reports on the fossil fuel wars of the Northern Cheyenne who,

> have been fighting off the [coal] mining companies since the early 1970s, in part due to an important Sweet Medicine prophecy that is often interpreted to mean that digging up the 'black rock' would bring on a kind of madness and the end of Cheyenne culture. But when I first visited the reserve in 2010, the region was in the throes of the fossil fuel frenzy, getting hit from every direction – and it wasn't clear how long the community's anti-coal forces were going to be able to hold out.
> (389)

18 *Introduction*

The appalling events that destroy ontologies and ecological relations with country go on and on. That there are so many 'disasters' has perhaps immured mining companies and their advocates to the damage caused by emissions of their mined resources. Already bad, what's more of the same? Western institutions complicit in the project of unrestrained economic growth and the commodification of everything are all at a crossroad. The museum is one of these institutions.

As I write in an arid and sparsely populated state with 'natural resources' spread across traditional lands of First Nations Peoples, the contradictions and enmities surrounding the extraction industries and ecology are in plain view. The Western Australian mining magnate Gina Rinehart is one of the world's wealthiest people with a net worth in 2019 of US$14.8 billion. Rinehart is a vocal supporter of expanding the fossil fuel and extractive industries. Her fossil capitalism is accepted by most Australians even as they know the power wielded by mining magnates and multinationals is not proper thinking. It is time for geo-inclusive museums to act, to enact kinship toward the Lithocene. As Jane Bennett (2010) explores in her seminal study on vibrant matter, we need new 'regimes of perception that enable us to consult nonhumans more closely, or to listen and respond more carefully to their outbreaks, objections, testimonies and propositions' (108).

Structure

Chapter 1 focuses on the disappearance of soils and the role that soil degradation and pollution plays in climate change. Museums conventionally display soils in ways that tend to downplay their teeming life and agency by a focus on pedometric markers and characteristics. This focus is ripe for change. Chapter 2 considers that greater critical nuance will expand the potential of museums to engage with debates on the efficacy of the Anthropocene thesis, as well as the concerns of posthumanism and the Posthumous. Museums can confront the utopian vision of trans-humanism and related techno-utopias that do little to dispel the nature and culture divide that needs to be superseded by entanglements of culturenature voicing the elemental operation of thinking.

Chapter 3 confronts the inclusion dilemma and considers ways to repurpose the museum to become a site that includes geological actants and agencies. This is not to disavow the significance of human diversity and access, but suggests there has been appropriation and commodification of inclusion by neoliberal rhetoric. In Chapter 4, climate fiction is considered as a way to illuminate the role that museums can perform in relation to the climate emergency. White Geology is considered in Chapter 5 as the instrumental use of the earth's materials and a force that geo-inclusive museums are in a position to counter. Chapters 6 and 7 consider the display of fossil fuels in museums and heritage sites. Coal and oil are evaluated as actants with agency that exceeds the way they are exploited and presented in modern culture. Chapter 8 conveys a role for museums in communicating modes of the geologic that are not bound by the colonial gaze. Chapter 9 extends the lens to gold and its puzzling lure across the centuries. It highlights the toxic impact of its discovery, processing and value on communities and environments as issues too often omitted in displays of gold as treasure, and in museums that present the history of a region's goldfield. The book ends by acknowledging the significance of museums as relevant sites of hope and transformation once the hubris of neutrality is shed.

1 Disappearing soils
Toward a pithier pedagogy

The importance of soil health to climate is not a widely or well told story. Currently soils remove about 25% of the planet's fossil fuel emissions each year, stored mostly as peat and permafrost. The depletion of these carbon-rich soils through erosion, drainage and cultivation releases significant quantities of CO_2 into the atmosphere (Lascelles 2015, 14). It is estimated that half of the world's top soils have gone in the past 150 years, and in 60 years the rest will disappear (Montag 2015, 30).

When soils are encountered in science museums and exhibits, they are pedagogic specimens understood through pedometric markers. In the cultural imaginary, dirt and soil are synonymous, and with the modern obsession with hygiene, soils are physically as well as intellectually voided from good, clean thought. Recent art does not shy away from the erotic and necrotic as wondrous qualities of dirt. Contemporary artists create perceptions that gesture to the powers of soil as inseparable from bodies, alive and dead. As a museum artefact, soils offer a unique material encounter.

Soil scientist Rattan Lal writes of 'an utter lack of awareness in the general public and among policymakers about the importance of soil and its sustainable management to national and international security' (Toland et al., 2019, 10). The soil receives less attention than mountaineering feats, the exotic underworlds of fiction and space travel. Given that life is supported by mycorrhizal fungi in a network of soil pathways about which very little is known, gardening should be the zeitgeist not Mars. The spirit of the age should herald extraordinary insights into fungi above the attention given to SpaceX.

The science of soil or pedology monitors changes in soil organic carbon stocks, as required by international conventions. Droughts, rainfall patterns and temperature all affect soil carbon storage. The data evidence soil erosion and disappearance; however, as with other aspects of climate, a focus on the facts does not necessarily convey the intimate entanglement of soil and climate in ways conducive to action. Museums can tell pithier stories of soil, of the inextricable entanglement of soil and humans with the warming earth.

Curator and environmental historian George Main (2017) believes that the inability to attend to destructive climate events on the land has to do with 'the

DOI: 10.4324/9780367741945-2

modern, industrial mind' not attending to 'our intimate, bodily ties to the rest of nature' (172). Main writes,

> Perhaps the challenge now facing museums, as the traumatic consequences of rapid, erratic changes in climate patterns begin to unfold, is to enable storytelling that constructs meanings that are themselves more powerful than the most ferocious storms, devastating fires, churning floods and bitter droughts.
>
> (177)

The power of a good story is evident here. Stories of soil are impactful by relatability; and responding to local and regional experience of communities contending with climate.

Main (2017) supports museological practices that offer ways to connect bodily to the land while 'maintaining intellectual rigour' (174). He cites The Paddock Report as such a project/story, an outreach website of the National Museum of Australia (NMA) that connects global climate issues with the everyday reality for a family farming a 60-hectare New South Wales paddock. The paddock is productive and resilient due to the ecological methods practiced by the family since the 1990s. Museums can foster such 'shadow places', Val Plumwood's term for the many remote places we overlook yet are ecologically and socially bound to (Main 174).

Agricultural practices contribute up to one third of the greenhouse gases responsible for climate change (Spaid 2019, 37). Main (2017) discerns that for museums like the NMA 'possibilities exist to reinterpret agricultural machinery and its history, to reveal and help undermine processes that generate climatic and ecological disorder' (117). He reinterprets a stump-jump plough, an invention that had a significant impact on land reform in the 1930s and 1940s, and that entered the NMA collection in 1987. The stump-jump plough, which enabled farmers to lift over tree roots and rocks to ease clearing of scrubland, tells the story that 'Agricultural scientists and technologists rejected the idea of adapting to local conditions' in rural Australia erasing Aboriginal knowledge of the country and its climate' (Main 2017, 116).

Pedology

In science museums and exhibits, visitors learn that soil is rock that has weathered at a particular place, in distinction to sediment, which is earth that has been transported from its origin to another site by water or wind. The stuff at the bottom of oceans is not soil or sediment but seabed. It was initially considered the role of museums to focus on the function of soils in the production of crops and livestock. The first soil museum, the V.V. Dokuchaev Central Museum of Pedology, was founded in Saint Petersburg in 1902 based on the extensive soil samples of Vasili Dokuchaev, who is considered the father of soil science (Kennedy 2015, 1134). In 2021, there were 38 dedicated soil museums identified in the world as well as 34 permanent exhibitions, 32 significant collections (accessible by appointment) and three virtual museums (Richer-de-Forges et al., 2021, 280).

Museums today are increasingly concerned with the role of soil in climate change mitigation. The relationship between soil and climate change links pedometric data and climate modelling, so science continues to be the primary lens through which museum visitors encounter soils. Conventionally soil is displayed as a monolith (a vertical slice of a soil extracted from the field) to reveal soil material in profile, 'a visual object that transmits to the general public a vision of changes in soil morphology with depth' (Richer-de-Forges et al., 2021, 294). Aesthetics creep into the science as, 'soil scientists tend to prefer to show amazing, distinctive soils' (301). Given there is already selectivity bias in the presentation of soils, there is potential to extend the aesthetic focus of the monolith display practice toward the astonishing micro life soils share with humans.

As a monolith, soil is observed with attention to colour, texture, functionality and location, characteristics that have historically enabled soils to be modified into an improved agricultural or other human resource. In the taxonomy of soil, location is a key characteristic, with few museums exhibiting soil monoliths beyond their own country or region. There is however a World Soil Museum at Wageningen University in the Netherlands. Wageningen is located in a large food production region and the university is known for its agricultural research. The museum has 1,100 soil profiles that have been sampled to a depth of 150 metres. Representing the 32 major soil groups of the world, 80 of the collection are on display. These soil monoliths are conserved through impregnation with diluted lacquer to a thickness of six cm (see https://wsm.isric.org/visit.html). The Wageningen museum website offers a virtual tour of the room where the monoliths are displayed. A map indicates where the samples are from, and a range of soil themes, including Soil and Climate Change, have embedded videos (https://wsm.isric.org/themes.html). This intriguing museum can expand the relation between soil and climate to highlight human entanglements not only through the micro level of living critters in soils but by incorporating leachates, plastics and plastiglomerate.

Most museums unsurprisingly display soils that are familiar to their region. Artist Margaret Boozer reveals the existence of local soil in her work *Correlation Drawing/Drawing Correlations* (2012). She extracted samples from the five New York boroughs and displayed these in plexiglass to show the correlation of soils with specific locations. The work gave a literal depth to the city, to the materiality of actual places. Perception shifts beyond the usual way of seeing the city as a surface concrete and asphalt infrastructure.

Many artists collect soils; they use pigments from soils in their work and sometimes soil is both the subject and medium of a visual practice. Dutch artist herman de vries has collected over 8,000 soil samples over 40 years, which are now stored in Musee Gassendi (Digne, France). Each sample is documented as a rubbing on paper, and these are compiled into artists books and earth museum catalogues. 'These earths, if necessary pulverized in a mortar, are strewn onto paper and rubbed up and down by his fingers'. The collections reflect that soil is specific to a particular place but de vries is also concerned to present individual soils vertically to block any association with landscape (Adams and Montag 2015, 42). For soil has an agency that does not correlate to the western landscape tradition of representation. A tradition that distracts from the perception that all matter, whether a tree or fungus, has an agentic force removed from human experience. Through his art practice de vries manifests an anti-sublime, an inhuman way of seeing, so we might differently encounter soil.

Humans-in-soil

Soils are a mash of dead plants and animals, living microorganisms, minerals, air and water. The mix of these materials provides a taxonomic way of ordering soil but this measure should not dull amazing intimacies and entanglements. There is little sense of humans *in* soil, no shared existence. Donna Haraway articulates a sense of becoming-with-soil as a physical/cognitive leap into natureculture. Her stories are a pithy play with dirt. 'Terrapolis is rich in world ... rich in com-post, inoculated against human exceptionalism but rich in humus, ripe for multispecies storytelling' (Haraway 2016, 11).

It goes mostly unremarked that people are compost, a neglect taken up by contemporary artists attentive to qualities of soil that remove physical thresholds with bodies. Sally Mann works with soil and decomposing corpses to document that 'we' are soil. Her photograph series *What Remains* (2003), for example, unambiguously engages with our mingled soil-ness in images of corpses laying on the earth as objects of study at the University of Tennessee's anthropological facility. That viewing the matter-of-fact images of becoming-earth feels like a type of voyeurism, focuses the viewer's attention on the modern distancing from death. Soil teems with death as well as life.

Artist Jae Rhim Lee's work is concerned to support human burial practices that revitalise the soil. There is increasing interest in eco-burials to which her work has no doubt contributed. Her ongoing *Infinity Burial Project* features a mushroom death suit that 'decomposes and remediates toxins in human tissue'. The burial suit is embedded with a special fungus that activates the practice of 'decompiculture' (http://infinityburialproject.com/mushroom). In bringing art and science together, Lee's project is rigorous and pragmatic, operating at a threshold that moves beyond cultural notions of the abject body. The project brings attention to the reality that industrial toxins are a part of bodies in the twenty-first century, and that death can be an opportunity for fungi to detoxify the soil of these toxins. The project promotes responsible engagement with death, decomposition and soil health and in doing so shifts attention to death as a future event of hope. Lee's Burial Suit, co-created with designer Daniel Silverstein, featured in the NTU Centre for Contemporary

Art in Singapore, as part of *The PostHuman City* exhibition (2020) (https://act.mit.edu/2019/11/jae-rhim-lee-featured-in-the-ntu).

Lee's creative practice exemplifies a mode of contemporary art that intervenes in the systems of the world. It supports William Fox's observation that the artistic and the scientific have 'co-evolved', that they attend to and reinforce each other. And that: 'Art museums, as it turns out, are excellent laboratories in which to test this proposition' (Fox 2017, 196).

Fixed disciplinary categories are not conductive to the keen attention of collaborative artists and scientists like Lee and Haraway, who value soil entanglements of *homo sapiens*. They do not regard the soiled human as an abjection but expose how the modern human has become cognitively detached from the body. Perhaps this explains fascination with crime fiction and forensic drama that lingers on corpses and processes of decomposition. Museums are sites to bring into focus the artificiality of an excessively clean way of thinking and to value soil perspectives by heeding the practices and teeming life of microbes, root systems and fungi as existents coevolved with 'us'.

Contemporary artists create erotic and necrotic soil perceptions that gesture to the power of soil as inseparable from bodies, alive and dead. Performance artists and filmmakers Beth Stephens and Annie Sprinkle enact eco-sexual soil weddings and play with dirt as a desirable entanglement. They combine wit and humour with the intense value of soil expressed as desire. Dirty Ecosexual Wedding to Soil was a 'full on wedding where two hundred and fifty people came and could all marry soil with us' (Stephens et al., 2019, 547). Over time their special rituals, such as ecosexual soil weddings, came to attract considerable attention. In their events, Stephens and Sprinkle are forthright about human responsibility for soils, but develop this without conveying blame.

> … we're politically very clear about what we're trying to do. We're really trying to shift this metaphor around the Earth from 'Earth as mother' to 'Earth as lover', to get people to engage in a more mutual way of being in and thinking about the world.
>
> (The Ryder Movies for Moderns)

Their play with desire is a potent way to communicate loss and the impact on local soils of a warming globe.

In museums of inventive composting, the drama is found in shallow underlands of aggregates, rubbles, pigments and burials. There are legends of philosopher's stones and fairy dust, but where are the tales of mysterious top soil, sexy sand and sentient sediment? Soils are distant from the glamour and dynamism accorded to dinosaurs, meteors and space travel. Thoroughly cleansed in the western cultural imagination the disappearance of soils is a hygienic subterfuge to avoid death. Museums are ripe to tap into visual soil languages where underlands are not fixed pedagogical maps awaiting industrial agriculture and Big Mining.

Dig it

Dig It! The Secrets of Soil (2008) at the National Museum of Natural History was a collaboration between the Smithsonian National Museum of Natural History and Soil Science Society America (SSSA). Having no previous museum experience, the SSSA assumed that once the importance of soil was explained, people would be impressed (Drohan et al., 2010, 703). On reflection, they acknowledge that this was not enough. 'If there is one key lesson learned from the entire experience of *Dig It!* it is that successfully communicating with the public requires inspiring the public' (704). The experience of staging the exhibition led SSSA to consider its own function, and to a new focus. They decided that soil taxonomy should have a digital presence so that a user, with no knowledge of soil, can use a graphical interface to locate soil in a landscape. The exhibition was considered a catalyst: '*Dig It!* is potentially the dawning of a new era for soil science and SSSA' (705). The organisers discerned value in branding their Society, 'As Levis is to jeans, SSSA is to soil science' (704).

While resolving to be bold and to think out-of-the-box, in evaluating the exhibit what appears overlooked are the pithy soil entanglements found in new materialism and contemporary art. Richer-de-Forges et al. (2021) write that soil inspiration requires a multidisciplinary dimension that involves soil science and art, in ways oriented toward communication and sharing (301). Moving an interface online so that interested people can identity their local soil is an example of translating disciplinary knowledge onto an online platform with the assumption that this transference is enough to garner a new kind of perception.

When science is involved with art, it is often to engage in cross-disciplinary projects that collaborate with on-the-ground, 'grassroots' work of local communities seeking to repair places in their environment or to prepare them for environmental disruptions. A number of such projects are documented in the anthology *Field to Palette* (Toland et al., 2019), a collection highlighting soil and art collaborations in the Anthropocene.

Cross-disciplinary engagement was also a feature in the survey exhibition *Deep Roots* (Falmouth Art Gallery, 2015 and Peninsula Art Gallery, Plymouth University, 2016). The exhibition and catalogue, like the anthology, document art practices that use science to engage with soils in ways distinct to standard agriculture and food production. While there are garden projects, particularly connected to schools, mostly the projects communicate alternative or experimental ways of perceiving soil and human relationships. For example, since the 1990s Mel Chin has collaborated with agronomist Rufus Chaney to explore the potential of plants that extract toxic metals from degraded industrial sites. Their pioneering Revival Field projects have advanced the use of hyperaccumulator plants in contaminated sites in the United States and Thailand (Adams and Montag 2015, 40).

Another artist with work documented in the anthology is Kenyan storyteller and filmmaker Wanuri Kahiu. Her film *Pumzi* (2009) imagines the world of the fictional Maitu Community who survive on a planet that has been devastated by drought through their use of intelligent soil (Rossee et al, 2019). The film has been labelled a postapocalyptic work of Afrofuturism (430); however, Kahiu points out that African culture has always used speculative fiction. The cultural role of seers or storytellers is to forecast environmental change. The film's central protagonist Asha is the curator of the Virtual Natural History Museum in Maitu Community. Asha leaves the museum after she comes to recognise that the most useful task and potential of a curator is to 'scrutinise, to question, and to investigate' the institution (441).

In thus framing a fictional African museum in a postapocalyptic scenario, Kahui draws on existing preservation archives like the Svalbard Global Seed Vault in Norway to provide 'a critique on society's interest in preserving things once they are dead or near extinction, while limited effort is made to keep them alive before they perish' (441). Here is a perspective akin to that of environmental philosopher Thom van Dooren (2017) who observes the dilemma that delaying the inevitable extinction of a critter by preserving them in 'arks' and 'banks' can actually provide an excuse for inaction. He considers whether a vague type of hope in keeping alive the last remaining members of a species whose world we have already destroyed is a kind of denial when 'what is needed is more attention and a critical lens of what it is we are specifically hoping and working towards' (152).

T.J. Demos (2016) examines ecological art through the lens of 'what it means today for an art exhibition to include ecological practices as art, and not just as models of radical gardening (242). He identifies a crisis in art in relation to climate and environmental disruption that he evaluates in the way that green 'garden-as-art' projects were presented in dOCUMENTA (13) in 2012. He observes that while gardens can be a vital way of attending to the financialisation of nature this context was not effectively made by the dOCUMENTA curatorial team. Demos finds a lack of curatorial discernment through the way Donna Haraway's speculative hybrid aesthetics in the exhibition profoundly diverged from that of Vandana Shiva's work, which addressed climate change activism wary of the type of biotechnical creations Haraway supports. Demos notes that not addressing the clash of positions presented in the work of these two important theorists,

> Risked a (non)-position of uncommitted theoretical pluralism, a tendency familiar in the liberal milieu of contemporary art, eager to allude to crises and emergencies – even to aestheticize them – but taking no clear stand in relation to them.
>
> (240)

It is an important critique of the exhibition, reflecting a form of neutrality when it comes to climate change and its disruptions. Again, to cite Demos, the exhibition 'is exemplary of the failure to take a stand about the very problems it raised – as if knowledge production releases us from any ethico-political responsibilities, or from assuming such responsibility at the curatorial level' (241).

Rubble

The *Field to Palette* anthology identifies six functions of soil explored through collaborative projects and dialogues – sustenance, repository, interface, home, heritage and stabiliser. The soil's stabilising function is a platform for human structures, infrastructures, and socioeconomic systems. The notion of soil as a stabiliser is ironic in relation to a number of the projects given the destructive impact of anthropocentric structures and the blip on the timeline of history that marks *homo sapiens*. In the context of geological time and duration, what is stabilised?

Irony is a tactic that can be effectively tapped to gesture to the enormous disruptions that climate poses and to inadequate efforts to mitigate and adapt to transformation. Stability measures – such as concrete, steel, taxonomies, data and institutions – are a temporary human foil, an awareness that is highlighted in the rubble installations of Spanish artist Lara Almarcegui, and the nuclear waste projects of the Center for Land Use Interpretation (2013; CLUI).

Almarcegui's (2015) installation *Aushub aus Basel* in Kunsthaus Baselland filled a large room with excavation matter from a demolished construction site (Almarcegui and Wessolek, 2019). There is a startling incongruity between the white cube gallery and pile of 'dirt' dumped into the pristine space. For the artist, the rubble from demolished buildings is testament to the obsolescence that attends modern design, and reveals the anthropogenic quality of urban soil. In its incarnation as construction rubble, the pile of material from which Almarcegui makes the work will eventually, over millions of years, weather and erode back to soil; it will exist forever in one form or another. It is impossible to comprehend the duration of this inhuman resilience; meanwhile the rubble in its late capitalist form says, 'look at what you have made me!'

We can approach inhuman matter and its agency to speak to us through various approaches, through the lens of contemporary art, or critical fields such as ruin archaeology and palliative curation, stances that engage with the anti-aesthetic dimension of modern ruins and discarded objects. Devalued materials – iron, glass, once-were-soils, synthetic fibres – constitute structures of modernity that are no longer stable. They offer an alternative focus to the attention given to heritage that is conceived as ancient or historical structures, sites that are fixed within the historical record, sites that are 'clean, fossilized and terminated' (Pétursdóttir and Olsen 2014, 7).

Critical archaeologists Þóra Pétursdóttir and Bjørnar Olsen draw attention to the otherness of less desirable ruins of the recent past by making decay or ruin itself, as with Almarcegui's rubble, the agent of attention. Such artefacts act as anomalies outside the order provided by hegemonic aesthetics, linear histories and heritage value. They encourage us to rethink the assumptions accorded to stories of order, continuity and progress that attend the modernist project.

> Being modern and ruined, made modern ruins ambiguous and even anachronistic, and their hybrid or uncanny state made them hard to negotiate within established cultural categories of waste and heritage, failure and progress. They became matter out of place – and out of time.
> (Pétursdóttir and Olsen 2014, 6)

Archaeological fieldwork on ruins of the recent past is a call for real, physical encounters with the material world, 'actual encounters with the masses of trivial, broken or soiled things gathering around us' (Pétursdóttir and Olsen 2014, 20). This expanded archaeology is beyond the anthropocentric value given to ruins by cultural institutions. A critical archaeology concerned with the way things assemble in their otherness is a significant value for museums to attend to. Directing attention to the way humanism limits how ruins and decay are understood in the western mindset, there is viability in a heritage perspective that embraces the un-useful. An approach, 'released from the imperative of domesticating things and ruins within the tropes of historical narration and identity building, and instead open to the possibility of appreciating them also in their otherness' (Pétursdóttir and Olsen 2014, 18).

In their field work and scholarship, Pétursdóttir and Olsen (2014) show archaeological excavation offering a new relationship with the recent ruins of human activity. The call for attention to recent things represents an adjustment away from the fetishisation of ancient ruins as a kind of 'finalised' heritage for what should be preserved and admired. They recognise that to confront abandoned and discarded objects of our own recent past is to acknowledge these objects' transient status; acknowledging this status of being in between and not belonging 'makes the ruins of the recent past so disturbing' (7). Acknowledging the recent 'ruins' of human endeavour draws attention to how humanism makes us comfortably numb to the presence of less palatable relations with materials.

> Abandonment, decay and ruination bring these relations to halt, they disrupt the routine and disclose things in their own unruly fashion, released from human censorship and order.
> (Pétursdóttir and Olsen 2014, 11)

Critical stances toward the discarding of modern materials use irony and documentation to expose the hubris of obsolescence, and in so doing can reveal when an endeavour has fallen dangerously short. In many of their projects, the CLUI adopt this stance. Their project *Perpetual Architecture* documents the underland storage of spent nuclear fuel in the United States. Founded in 1994, CLUI is a nonprofit research organisation based in California, 'dedicated to the increase and diffusion of knowledge about how the nation's [USA] lands are apportioned, utilized and perceived' (https://clui.org/section/about-cen ter). *Perpetual Architecture* observes the use of land for radioactive disposal sites. The prospect of a human architecture that is 'perpetual' is ironic as the sites are tasked with containing the radiation from uranium 238, 'which has a half-life of 4.47 billion years, nearly the age of the Earth itself' (Toland et al., 2019, 575).

The nuclear waste burial mounds are located from Pennsylvania to Arizona (see www.clui.org/page/radioactive-disposal-sites-usa). From the air, the sites are large, low flat sandy mounds, with the outer shell covering the buried material. Mostly built on arid land these mounds or cells contain radioactive tailings and other material from uranium processing. The U.S. Department of Energy aims to make the sites not prominent; however, the attempt to make them appear part of the landscape only makes their presence all the more sinister. 'These unintended "land art" objects represent some of the most menacing surface features on the soil platform' (Toland et al., 574). The Center evaluates that the mound sites

> are the end of the line, meant to be unconnected to the rest of the world, like deadly anachronistic time capsules. These are the most negative of spaces, nonplaces, meant to stay inert and isolated for as much of forever as possible, kept from the present, but destined for the future.
> (CLUI 2013, 239)

CLUI's terminal atomic mound project is fittingly the final entry in an insightful collection of projects *Making the Geologic Now* (Ellsworth and Kruse 2013). The entry is preceded by a remarkable 2009 interview with a U.S. Department of Energy geoscientist on the design of future nuclear waste facilities, particularly a proposed project at Yucca Mountain, Nevada. The idea is to construct a deep facility that will remain extant for about a million years. It is an act of engineering and geological ingenuity, with the waste to be buried in containers that will fill long caverns or emplacement drifts. There will be 91 of these tunnels into the mountain, each with 120 waste packages. The depository would be in use for around 100 years, and then sealed. Yucca Mountain is a relatively stable and arid site so may last a million years without a damaging earthquake or water leaking into the containers, either of which could compromise the situation. As the geoscientist observes, many countries are researching similar facilities. Museums for the end of the world.

The entanglement of earth with waste materials produced by *homo sapiens* expands year by year. Not only are the remains of humans alive after 1945 forensically datable from the fallout of atomic material, but mountains, deserts and other underground environments are being stuffed with hidden toxic waste. That we are somehow separate from the material realm of soils and underlands is a delusion; the truth is, as Jane Bennett (2013) reminds us, that we Earthlings, are 'walking, talking minerals' (244).

The degradation of soils includes transformation into 'waste' agents, which is a process of material change enabled because the waste is understood as imperative to economic growth. Myra Hird (2017) studies the actant created in waste depositories known as leachate: 'a heterogenous mix of heavy metals, endocrine-disrupting chemicals, phthalates, herbicides, pesticides, and various gases including methane, carbon dioxide, carbon monoxide, hydrogen, oxygen, nitrogen, and hydrogen sulfide' (260). Hird uses the way this actant operates to caution against an Anthropocene aesthetic that 'relies upon a Western Enlightenment ideology of mastery and control over our environment and ourselves and [finds] salvation through increased reliance on science, engineering and technology' (Weinstein and Colebrook 2017, 255).

This aesthetic explains why the pollution of urbanisation, such as leachate, signifies power and progress rather than catastrophe. In the modern ideology of mastery over nature, the innovations of science and technology can fix any problem encountered on the road to progress. Hird concludes however that waste has actually become a force that exceeds the aesthetic of the Anthropocene. The assumption that the waste of rampant capitalist consumerism can be controlled by the rational order of science and technology is wrong. Soils are degraded and transformed by the materials that constitute modern landfills. The list includes diapers, metals, liquids, refrigerators, pet shit and litter, dead pets, batteries, food and fabrics, as well as products of industrial processes, over seven million known chemicals and 14,000 known food additives (Hird 2017, 260). These sites assemble billions of bacteria. Deep in the landfill strata is where anaerobic bacteria produce leachate. We like to think that leachate in waste dumps won't leak, just as we expect nuclear waste to be contained in burial mounds. The leachate provides a novel challenge to bacterial communities, which always 'figure out ways of metabolizing whatever matter energy they encounter' (261). The conceit that somehow it will be sorted out by human know-how means we do not see our waste or its unintended consequences, which Hird muses, 'comprise one of the deepest and darkest ecologies of the current material-historical juncture' (261).

Geo-tubing

An initiative documented in *Making the Geologic Now* is the Dredge Research Collaborative (2013; Stephen Becker, Rob Holmes, Tim Maly and Brett Milligan). The researchers offer a valuable perspective on the way that the relatively recent technology of geo-tubing impacts and enables a global dredge cycle – a monumental reconfiguration of dirt as a type of landscape design. They observe that the ubiquity of geo-tubes goes largely unnoticed as architecture – they are an engineering and logistics function without an aesthetic focus – yet these mushy wormlike tubes increasingly shape the landscape.

> A geo-tube is like an oversized sausage casing made of 'geotextile', a synthetic fabric woven primarily from hydrocarbon-derived polymers. When

deployed, they are inflated by liquids, slurries, or sediments, depending on their intended use. Geo-tubes find their application where water meets land and where landscape meets industry. They are deployed along riverbanks, coastlines, in shallows, or wetlands. They have spread quickly thanks to their flexibility, speed of application, and cheapness.

(Ellsworth and Kruse 2013, 72)

These geo-tubes are used everywhere, enabling the movement of wet material including sand, silt, clay and water. This liminal engineered geology shapes landscapes through moving sediment, a post-natural impact described as 'entirely appropriate for an era in which we are freezing sediment-spraying rivers' and 'impounding the eroded sediments of entire continents behind vast concrete structures, like Three Gorges Dam. Our largest monuments are not pyramids and skyscrapers but geologic impacts' (77). Terrifying.

The modern compulsion to order the Earth is a function familiar to museums. They are good at it. It is an authority and trust that can be repurposed. There is much territory to uncover once the cognitive plunge is made into entangled systems of inhuman coevolution. It is evidentially wrong to perceive soil as static and inert, given its multitude of species (most of which remain unidentified in the Linnean sense). 'A teaspoon of soil supports more organisms than there are people on this planet. A single gram can contain several billion bacteria from thousands of different species' (Adams and Montag 2015, 13). There is increasing awareness of the human biome, but soil is 'the mother of all biomes' acting as 'a vast digestive system, the collective stomach of all terrestrial plants' (2015, 8).

As technology enables the observation of ever smaller particles, awareness expands of the dynamic multitude and enigma of soil microbiomes. The recent observational capacity to see smaller things contributes to critiques of scale in the environmental humanities that bring agency to actants previously outside the scale of human perception. Timothy Clark considers that many forms of environmentalism are loose manifestations of scale critique (Fritsch et al., 2018, 85). While these scale critiques have a certain spectacle that might be applied to the way soil is exhibited, this does not necessarily translate to change in thinking. Yet the spectacle of the minute and once-mundane is an affecting tool, as David Attenborough's popularity affirms. The impact of Attenborough's personal closeness to 'the other' is yet to be fully understood, but he has enabled storytelling of the miniscule and immense, such as geologist Jan Zalasiewicz's docu-fiction *The Planet in a Pebble* (2010).

The story of a pebble is a scale critique describing the life (geology, chemistry and biology) of a single pebble from the origin of atoms in the Big Bang, through billions of years of underground and surface existence on Earth to its form as a slate pebble located on a Welsh beach. Zalasiewicz extends the pebble's journey into a future beyond human extinction and explains the processes as the pebble returns into elements and atoms. In a post-Earth space as cosmic dust, it drifts across the galaxy. Zalasiewicz's study is a reminder of our staggering irrelevance as a species, yet a species able to imagine the life of an inhuman other.

Soil(ed) words

In many respects, words fail soil. Dirt and soil are indistinguishable terms yet dirt is waste, and we cover waste, hide it underground, make it invisible. It even lacks the lustre of 'dust'. Compared to dust, soil lacks poetry. Dust 'is the stuff of fairy tales, stories of deserted places – of attics and dunes, of places from so long ago they seem never to have existed' (Parikka 2015, 85). Soils are just soils.

'Soil' was not always 'soiled'. It comes from the Latin words *soilium* meaning 'seat' and *solum* meaning 'ground'. How did soil come to be associated with contamination? The word *sol* or 'earth' refers to the ground, but not specifically 'dirt'. The ground, though, is devalued by the cultural value attached to height. The ground is a lower, lesser place than heaven with high and low, 'strongly charged words in most languages' (Tuan 2014, 34). Whatever is superior or excellent is elevated, and so associated with a sense of physical height. '"Superior" is derived from the Latin word for "high"' (34).

The early meaning of earth, 'from the Old English terms '*ear(the) e* and *ertha*, and the German *erde*' were in usage before the concept of Earth as the blue planet imaged from space (Montag 2015, 20), an event that transformed the human sense of itself as a species. The way that many people 'know' the Earth comes from photos of the planet seen from space. With this event of distancing, from the great 'height' of outer space, soil is nothing. Knowledge zooms from the surface upward into space, into the heavens. The soils of other planets are up there, waiting to be mined. We can imagine our globe, our planet, yet do not see our soil. And all the while the impression of earth from space is illusion. The most common photo of earth from space, outlining the contours of North America, is actually a composite image that Nicholas Mirzoeff (2015) observes is 'a good metaphor for how the world is visualized today. We assemble a world from pieces, assuming that what we see is both coherent and equivalent to reality. Until we discover it is not' (10).

As well as 'not the earth', soils are simplified as symbols that do not measure our complex entanglement with them. The familiar symbol for soil is a hand holding a piece of soil with a plant growing from it. This is a universal symbol; it was used to promote the United Nations 2015 Year of Soils. But it is an anthropocentric image conveying that mankind has the soil in hand, under cultivation, under control. An idea that is not something to gloat about given soil degradation and loss with the automation of agriculture and climate change. Soil scientist Edward Landa wants the simplistic imagery to change, 'In terms of images – please, no more cupped hands with dark, rich soil and the optional seedling. Give us more to chew on!' (Toland et al., 2019, 314).

Underlands

In his work *Underland* (2019), travel writer Robert Macfarlane takes the reader through caves, mines and catacombs to reveal 'how resistant the underland remains to our usual forms of seeing; how it hides so much from us, even in our age of hyper-visibility and ultra-scrutiny. Just a few inches of soil is enough to keep startling secrets …' (100).

Soil does not have the cultural presence in the western imaginary that is given to the deep underground, which is attached to romantic 'underworlds' where heroes and mighty beasts roam. Soil is mundane rather than wondrous. The deep underworld is an environment from which nature has been effectively banished (Williams 2008). As such these underworlds are not dirty or abject as are soils. Soils are a nowhere between the deep exotic underworld and bright human surface. Far beneath the grubby soil is where we find mythical journeys to lost civilisations and cities, and the centre of the earth. An underworld sublime operated as a powerful perception between 1700 and 1900:

> The sublime underworld was neither ugly nor beautiful, but something else entirely; obscure but pleasingly obscure, terrible but delightfully so. Gradually, however the underworld came to be perceived as wholly beautiful – as an illuminated artificial paradise, a splendid refuge from nature's imperfections.
>
> (Williams 2008, 83)

Soils are not sublime; they are abject. Abjection makes it easier to transform soils into a technology to engineer for human use. We make soil fertile with pesticides, and hide our waste and poison in it. What perfidy! Soil-human actants as complex entities with myriad ways of existing are far removed from the usual distancing of human cognition from the ground. Soil *as us* is a radical alterity that requires pushing the imagination to acknowledge a Terrestrial mired in experiences conceived like Haraway's Chthuluscene.

Soil-humans do not sit comfortably with the Anthropocene, one beginning of which is placed with the intensification of agriculture around 2000 years ago. Like much about the age of the geological impact of man, this beginning is disputed. Farming soils began long before this, an event pedologist Ronald Amundson calls 'the most unnatural act imaginable to our planet'. Farming removes the existing ecosystem and physically disrupts the upper layers of the land surface on a continual basis. 'The resulting erosion, CO_2 emissions, and biodiversity loss by farming are staggering' (Toland et al., 2019, 73). Like the image of the cupped hand holding soil and seedling, the perception that farms and crops are 'natural' is simplistic.

There is nothing simple about the remarkable underground of fungi. The kingdom of the grey suggests a metaphor for a reformatted museum of alterity, of altered thought, of absolute difference from what humans think they are. 'All taxonomies crumble, but fungi leave many of our fundamental categories in ruin' (Macfarlane 2019, 102). Mycorrhizal underlands are still so little understood they escape the binary opposition of nature and culture that fixates how modern subjects know the earth. Fungi is the super realm of alterity; its utter otherness to the way that the organic and inorganic are categorised by science offers a challenge to usual models of time, space and species (2019, 101).

Fungi is only recently a serious actant in western science where its agency and autonomy defy categories of knowledge. Biologist Merlin Sheldrake's engagement with fungi outlined in his study *Entangled Life* (2020) is science communicating thresholds that stretch the imagination. Sheldrake and other plant people and explorers of the 'wood wide web' provide a foundation to deal with the transformation of thinking that is necessary to address environmental catastrophe. Sheldrake explains how lichens have been a gateway organism for scientists to apprehend the reality of symbiosis, forms that confuse 'our concept of identity and force us to question where one organism stops and another begins' (80). His study of mycorrhizal relations finds that 'we might dance to their tune more often that we realise' (20). Here is inspiration and metaphor for wild, mushroomy museums that challenge how people know the world. 'You look at the [mycorrhizal] network…and then it starts to look back at you' says Sheldrake (cited in Macfarlane 2019, 100).

Sheldrake (2020) shares his experience at the edges of 'acceptable' biology, and reflects that the scepticism he encounters in his research arises because scientists must appear credible in accordance with extant knowledge schemas. His experience is that 'imagination usually goes by the name of speculation and is treated with some suspicion' (21). Here is an opening for museums to harness the capacity to take imaginative leaps and take visitors into the entangled places of thought where assumed thresholds between species and existents no longer hold true. Truly perceiving interconnection with things can be mind-blowing. As Macfarlane (2019) muses, 'Occasionally – once or twice in a lifetime if you are lucky – you encounter an idea so powerful in its implications that it unsettles the ground you walk on' (87).

2 Sinking and melting
Glossing the climate problem

For museum futurists Diane Drubay and Asha Singhal (2020), the climate crisis requires that museums work toward transformations that will have an impact at the level of systems. 'The power of resilience [to climate] lies in emergent cooperative systems, both within and between species' (2020, 665). Museums can support thinking that enables 'cooperative systems' between species. The analogy they discern between how museums can practice as an interconnected operation and the way nature works is itself a challenge to usual understandings of how human systems work; it is 'timely and necessary to understand how to apply the rules of Nature to the world of museums' (668). In nature it is the local that is the starting point; there is no central command; 'often smaller local interactions impact the global design of larger systems', be this fireflies or slimy molds (ibid).

What is curious is that this communication is not far advanced in the museum sector. Critical thinking across disciplines has long known that 'nature' as a separate entity to 'culture' is itself a cultural judgement. Alternative theories of entangled natureculture, whether or not this term is used, are the very basis of recent areas of research be this evolutionary biology, ecological art history or posthumous philosophy. Compelling ways to stretch the imagination and to create new objects of thought are supported by researchers recognising that the assumed thresholds between things are far more complex than standard binary thinking.

Weathering the Anthropocene thesis

Displays on climate are not uncommon in science museums, and representing the weather continues to be a significant theme for artists working in the landscape tradition. Recently, however, the focus on climate is framed by the Anthropocene thesis – the proposition that *homo sapiens* has altered the geology of the planet. How the thesis has been adopted and adapted by museums is informative in relation to the perpetuation of the culture and nature divide.

Anthropologist and geographer Lotte Isager and colleagues (2021) surveyed 41 exhibitions held from 2011 to 2019 in art museums, cultural history museums and natural history museums dealing with the Anthropocene. Of these exhibits 33 were held at art institutions; the earliest, *Anthropocene Extinction* by American street artist Swoon, was held at Boston's Institute of Contemporary Art. What the researchers conclude from their analysis of the exhibitions is that overwhelmingly there is no clear stance to act now on climate change. Instead, the exhibits involve 'open questions, uncertainty and the lack of settled truths' (Isager et al., 2021, 96). The focus for museums is with making the topic 'accessible and relevant for a general audience' (96). The research identifies that a difficulty with this stance is lack of specificity glossing over the central problem that what must end is burning fossil fuels and, relatedly, the structures underpinning economies of limitless growth. In confusing the issue by a lack of specificity, museums remain neutral, keeping issues conversational even though the time for debate over the science has come and gone.

A method the researchers observe is that museums often cluster together objects and subjects that were once considered to come from disparate realms, with the intention to 'establish a new sense of reality'. They question the effect of this practice asking how 'flat' these realities actually are 'in the ways they portray natural-cultural assemblages' (Isager et al., 2021, 99). While museums might recognise the need to challenge the 'Enlightenment's ontological divide between nature and culture' and even accept that this challenge remains the core of most museums' vision of their own role as museums in an Anthropocene world (Isager et al., 2021, 97), this falls down in practice.

The problem is that while exhibits are intended to present ontological connectivity in the portrayal of nature and culture, humans continue to be central, whether as a ' "villain" (the white industrial colonizer of nature) or "potential saviour" (the white scientist/technological innovator)' (Isager et al., 2021, 99). What the incongruence between acknowledging connectivity and yet displaying human sovereignty suggests is a universalising that is habitual in the museum. Controversy is avoided. The 'exhibitions put on show an Anthropocene and a "future-assembling" that deliberately exclude the controversies about both the concept and the predicament of the world from their arenas for reflection' (99). The conclusion is that,

> By downplaying the controversies of the Anthropocene, many exhibitions dealt with in our analysis do not appear to have opted for what we believe is the right conversations about this topic but, rather, perhaps simply the conversations that were deemed possible to have. If the concept of the Anthropocene, as critics argue, implies a large degree of whitewashing of historical inequality and exploitation, museums may inadvertently exacerbate the very problems addressed in the exhibitions by choosing to communicate with their audiences through this term.
> (Isager et al., 2021, 100)

The insight accentuates museum governance that adopts a stance of neutrality. Museums are not challenging basic inequities, including the historical concept of nature that is their own established legacy. Museum representation of nature with a capital N communicates the wrong message when it comes to real action to mitigate climate change. An effect of the Nature metaphor is to categorise all non-humans and earth processes as the other, so that an awareness such as Merlin Sheldrake's that humans may dance to the tune of mycorrhizal networks cannot be processed as logically plausible. As Fiona Cameron (2017b) writes,

> The nonhuman, animals, insects, rocks and earthly processes are therefore relegated to the position of a passive object to be documented, described, named, protected, controlled and used for human ends.
>
> (51)

Cameron's liquid museum as a response to responsible action toward climate futures provides an exemplar for an expanded museum of inclusion. As her research reveals, 'Curatorial exhibition, collecting, documentation methods and practices are still characterized by unrelenting dualisms and substance metaphysics with their corresponding sense of a subjective mastery over nature …' (2017a, 19). Her reading of the 2010 exhibition *Atmosphere: Exploring Climate Science* (at the Science Museum, London) discussed ahead evaluates how text and image reiterates the exclusionary separation of humans and non-humans that functions to distance 'us' from climate. Removing the human from the climate narrative, Cameron argues, fails to adequately acknowledge our entanglement with the non-human world together as agentic forces (2017b, 61).

Out of our hands

A complex view of nature and culture was sought by the large-scale 2014–2016 cultural history exhibition *Welcome to the Anthropocene: The Earth in Our Hands* hosted at the Deutsches Museum (Museum of Science and Technology) in Munich. The exhibition was a collaboration with the Rachel Carson Center for Environment and Society, which initiated the exhibit as a form of public outreach (Þórsson 2020, 105). It was an ambitious endeavour and it is timely to consider its success in challenging thinking around museum governance of climate change.

In planning the show, it was felt the museum's science collections were of primary importance as technical measuring devices and other objects 'provided suitable real display items for explaining the technical and scientific dimensions of climate change and other biophysical changes at the end of the Holocene' (Robin et al., 2017, 254). The idea was to avoid using these collection objects to convey a message about the linear advance of technology. The organisers adopted the concept of 'usworld/usweld' to present an imaginary that 'blends nature, culture, technology and society into [a] single hybridized perspective' (Robin et al., 2014, 211). The 'message' of the exhibit was intentionally not a narrative of decline, but rather aimed to present feedback loops that reflected objects as integrated and hybrid systems (212). The narrative that was ultimately presented did emphasise the consequences of resource management and exploitation in the era of industrialisation (Isager et al., 2021, 97) but the exhibition does not dwell on the possibility of dystopian futures.

The exhibition organisers were aware of the complexity of the topic and that, 'it is quite difficult to build the Anthropocene around museum objects' (Keogh and Möllers 2017, 87). They reflected that perhaps the difficulty is actually an aspect of the message to be communicated, 'The Anthropocene exhibition may be a first step in realizing that our world is no longer in order' (87). Clearly an attempt was made to present a different kind of visitor relationship with objects to communicate this realisation. The shift from presenting a world of order to an epoch of instability was manifest by 'nonlinear opportunities', such as visitors choosing their own path through the space (Robin et al., 2014, 213). But it does not follow that the expression of uncertainty was particularly impactful. Perhaps because the Anthropocene names an era and in doing so makes a human order of the world. In this, it is performative, that is, it creates what it names. So to utilise this order or category without a clear message that order itself is an anthropogenic impact is inadvertently disingenuous.

The intent was for visitors to experience the Anthropocene in such a way that ignited their sense of participation in the exhibition (Isager et al., 2021, 96). It is difficult to evaluate the effect of curatorial tactics although while clearly keen to address the geological (through use of the term Anthropocene and six display platforms representing Earth's tectonic plates), the choice to not highlight the complex networks of actants that comprise the climate missed an opportunity for a more embodied, present and intimate participation. Had the exhibit been located on the sinking island of Tuvalu, or a transplanted Inuit community or in New Orleans, the damaged earth's penetration as a narrative may have been franker.

Isager et al. (2021) conclude that the Munich exhibition's focus on the Anthropocene, along with most of the 41 Anthropocene exhibits they analyse, has the characteristics of a topic that is uncertain and hard to pin down, and attached to a cluster of topics – climate change, ecological crisis and industrialisation. As a consequence, 'rather than a well-defined issue, the Anthropocene is displayed as a concept in motion' (98) that bypasses controversial topics and is not specific 'about the culprits and reasons behind the environmental crisis' (99).

Bergsveinn Þórsson (2020) conveys his experience of the exhibition through a self-guided tour of *Welcome to the Anthropocene*. The lure to a more complex interaction with the topic happened for him with a final object he encountered. The object was a monkey wrench, a largely obsolete hand tool that was not originally part of the exhibit, but part of an Anthropocene Slam added after the exhibition opened (114). The Slam was a cross-disciplinary event that considered what a Cabinet of Curiosities for the age of humans might look like and include. 'In this era of extreme hydrocarbon extraction, extreme weather, and extreme economic disparity', the Slam invited debate and reflection on how certain objects might 'make visible the uneven interplay of economic, material, and social forces that shape the relationships among human and nonhuman beings' (Mitman et al. 2018, x). In 2015, after the official exhibition was opened, objects from the Slam were added.

For Þórsson the monkey wrench acted as a slippage of the Anthropocene concept and in doing so captured the absence of a deeper critique. I am reminded in this 'slippage' of Kevin Hetherington's discussion of the impact of a slipware jug, made popular on a UK TV antique show and affectionately named 'Ozzy the Owl'. When the jug was displayed in a UK museum, following its somewhat reluctant acquisition, it came to outshine for visitors the more high-brow ceramics (Hetherington 1997). The object disrupted the museum's aesthetic and legacy of connoisseurship; it gave the Wedgewood collection the 'slip'. The impact of an object whose inclusion lies in the fact that it is an anomaly should not be underestimated.

The monkey wrench is a counter cultural symbol for protest against the destruction of wilderness and the environment. In 1975 Edward Abbey published *The Monkey Wrench Gang*, a novel about a group who use sabotage to protest against environmental destruction. The book cover illustrates a wrench with a bulldozer pincered between its jaws. The tool was adopted by the environmental advocacy group Earth First in 1980; their logo is a wrench and stone hammer. More recently, a wrench is an object in the 'Antarctica Collection', part of artist Amy Balkin's exhibition of collected items, *A People's Archive of Sinking and Melting* (2012).

Once a valued hand tool, the wrench is now largely redundant, which contributes to the way it performs as a lure. In an essay on the wrench, Daegan Miller (2018) writes, 'once used everywhere lithe human muscle struggled against iron intransigence, the monkey wrench had a hand in building the entire towering, not tottering mechanical skeleton of the industrialised modern world' (144). Miller says the tool functions to twist the standard gauge of linear history into a question mark. A tool of human labour, the wrench offers us pause to consider whose labour built the Anthropocene, and who profited. Miller (2018) owns a wrench and paused to hear its perspective.

> I listen again, and realize that the monkey wrench's greatest strength – indeed, its intended purpose – is to turn the bolts connecting dissimilar things. Perhaps in the Anthropocene, the wrench has a newfound purpose: securely bolting nature and society – whose separation has long signified the triumph of the modern, capital-hungry world – back together.
> (147)

Techno-utopias

Separating non-humans and humans is an artificial act performed through disciplinary knowledge that is based on philosophy of difference. The reality that things are always muddled entanglements is masked so that a unique entity of man can be defined, categorised and slotted into the order of things. Although this philosophical dualism and its logic is widely recognised as flawed, it persists in everyday common-sense approaches to the world (Shaviro 2014, 152). We think in dualisms without being aware that this is not the material experience of our movement with the physical objects and systems of the earth. It is apparent to many that the status given to human separation from material processes and events is changing, as anthropologist Philippe Descola (2013) writes: 'One does not have to be a great seer to predict that the relationship between humans and nature will, in all probability, be the most important question of the present century' (18). That the tenacious logic of difference persists in museums supports the contention that critical museology is an underdeveloped field and goes some way to explain Robert Janes' and others' frustration at the global lack of museum activism on climate change.

Critical depth is important to distinguish whether a posthuman theory or idea is an extension of dualistic thinking rather than the inhuman 'death of the anthropocentric world' that is required. To reach a theoretical stance that does not centre the human involves acknowledging that the 'human' is not what humans think it is. Cary Wolfe argues that this 'human', this 'humanity' has never existed (Beckman 2017, 48). This does not herald a new idea. Foucault noted in the 1960s that what we know as 'man' was the effect of a change in the fundamental arrangements of knowledge. 'We' are a recent arrival: 'As the archeology of our thought easily shows, man is an invention of recent date. And one perhaps nearing its end' (Foucault [1966] 1989, 422). What these scholars help clarify is that the 'human' is a staged affair performed through practices and institutions of western humanism, and in this the museum has been instrumental.

Much of what we encounter as posthuman in museums and elsewhere refigures a utopian belief in the power of science and technology. On this, Wolfe critiques philosopher Nick Bostrom's trans-humanism as an intensified humanism that derives 'directly from ideals of human perfectibility, rationality, and agency inherited from Renaissance humanism and the Enlightenment' (Wolfe 2010 xiii). There is a difficulty, Wolfe notes, in not differentiating between humanism and Enlightenment thought. Humanism is a universal dogma while Enlightenment is a process of using your own understanding. The task for Wolfe, and that I extend to museums and inclusion, is what thinking has to become to confront the challenges of climate and environmental crises. In this, the difficulty is not humanism per se (which has values and aspirations to admire) but how the frameworks used by humanism 'reproduce the very kind of normative subjectivity – a specific concept of the human – that grounds discrimination ... in the first place' (Wolfe xvii). In this context, any 'post' human museum needs to be an instrument of enlightenment, to think for itself and lure visitors by its imaginative engagement with new knowledge, to prepare a materialism for our time and to make confident futures beyond fossil capital.

Optimism in human authority and agency to control earth systems is encouraged in notions of enhancing humans beyond the current limits of life. Trans-humanism or intensified humanism champions the faculty of reason over the physical body, as if they are distinct. Bostrom is a key figure in this version of posthumanism setting up the World Transhumanist Association known as Humanity+ www.humanityplus.org/). He promotes that 'current human nature is improvable through the use of applied science and other rational methods' (Nayar 2014, 6).

Weinstein and Colebrook (2017) discern that to be transhuman is to 'conquer death' through enhancement that 'does little to advance the human beyond itself and in that sense is not really "post" anything' (xv). The most popular versions of posthuman enhancement push toward breaking with the organic evolution of the human species, a stance earlier promoted by Ray Kurzweil and Hans Moravec, 'who believe that digital technology in various forms will emancipate humans from their bodies by merging human consciousness (in particular rational intelligence) with machine software (robots, for instance), thus enabling humans to transcend the limitations of their bodies' (Anderson 2017, 18). When this moment of singularity is reached, the flesh and biological roots of the human will be irrelevant.

Yet in extreme weather, now, it is the body that starves, burns and drowns.

What the hope of singularity conveys is that the challenge to escape the limits of humanism is complicated, '... where and how to position posthumanism in relation to humanism is clearly fraught with difficulties and disagreements' (Beckman 2017, 47). As well as rejected by thinkers seeking to move beyond dualistic perspectives, trans-humanism relies on a faculty of reason that recent understanding of brain functioning refutes.

> Real reason is: mostly unconscious (98%); requires emotion; uses the 'logic' of frames, metaphors, and narratives; is physical (in brain circuitry); and varies considerably, as frames vary.
> (Boulton 2016, 773)

In other words, reasoning is a largely unconscious habit of framing information into recognisable form. Posthuman reasoning framed through new technologies dictating the rate and trajectory of change is a frightening prospect, which is not to concede to an anti-modernity that rejects advances in technologies that cure terrible diseases or forward sustainable energy sources.

There is an overarching transcendent thinking behind an overly keen avowal of human knowledge that overlooks that it is this knowledge that has produced 'the ravages of both a savagely unjust economic system and a destabilized climate system' (Klein 2015, 8). Humanist/post-humanist techno-utopian fantasies can be repurposed by museums, which will then play a significant role in enabling visitors to notice that a hypermodern future as a desirable goal is misguided. Rather than promoting an updated human, museums can present a modest anti-humanism with perspectives by the earth, which is surely what the climate is articulating.

Atmospherics

Fiona Cameron (2017b) provides a critique, and an alternative approach to the exhibition *Atmosphere: Exploring Climate Science* that opened in 2010 at the Science Museum in London. It is useful to note her views on the hubristic messages communicated by the exhibit, a messaging she observes is iterated in other displays on the atmosphere such as *Climate Change Wall* (National History Museum, London), *Ecologic* (Powerhouse Museum, Sydney) and *Dynamic Earth* (Melbourne Museum) (Cameron 2017a, 20). These displays and the Science Museum exhibition are inflected around the message that climate can be fixed by science and technology. There is a mismatch between the reality of climate as an unpredictable force and the way science is the tool to control future climate scenarios. A message this imparts is that nature is out of sync and 'if properly managed through a series of technological, managerial, and organisational solutions, then life can be securitized again' (Cameron 2017b, 70).

Several years prior to *Atmosphere*, the Science Museum held *The Science of Survival – Your Planet Needs You* (2008), an interactive display encouraging visitors to use the provided data from climate modelling to design a sustainable future city. The exhibit is considered fairly typical of the way climate change is presented in science museums and centres. In focusing on data to project a distant future scenario and in the format of its presentation as a one-off campaign, it is described as 'failing to generate a deep sense of change of consciousness and practice' (Salazar 2017, 101).

The tendency to not engage with the contradictions and complexities of climate also occurs when an exhibit puts the responsibility for social change onto individual visitors. Cameron observes this in the 2009 exhibition *Climate Change: Our Future, Our Choice* at the Australian Museum, Sydney, which deployed, 'A behavioural psychological mitigation imaginary disciplining the responsible individual to Do Something to reduce their carbon footprint to secure a precarious future as opposed to Doing Nothing, leading to inevitable climate catastrophe' (cited in Salazar 2017, 95). It is increasingly acknowledged that placing blame on individual domestic behaviour for global crises is ineffective. Rebecca Huntley (2020) writes, 'the prevailing view is that trying to make people feel guilty and therefore responsible can be a risky strategy. It can lead to anger, resistance and avoidance, and can discourage deep listening and understanding' (80).

Jeff Sparrow's research reveals that making consumers feel guilty is a well-known tactic of polluting companies. One of his examples is the Keep America Beautiful (KAB) campaign, which was a deceptive initiative to encourage recycling that was based on generating consumer guilt.

> By the 1980s, [KAB] was receiving millions of dollars each year from 200 companies collectively responsible for one-third of the material ending up in American landfill. Few of the ordinary Americans who supported its work knew that KAB was funded by corporations that incinerate toxic waste, nor that its directors included representatives of Philip Morris, Mobil Chemical, Procter & Gamble and PR giant Burson-Marsteller.
> (Sparrow 2021, 103)

In her analysis of the didactic texts in the *Atmospherics* exhibit, Cameron discerns a specific meaning around science and technology as the solution to climate change. Here is one such text:

> Tackling global warming by deliberately manipulating the Earth's system will rely on finding practical, affordable methods for capturing carbon dioxide. This technology could let us keep burning coal and gas but without most of the carbon dioxide emissions.
> (Cameron 2017b, 63)

The myth of stabilisation that Cameron (2017b) concludes is a central message of the exhibition, is also exemplified in 'hubristic interactive games' that support 'the modern project' by locating straight forward linear solutions to climate change such as challenging participants to cut emissions by 2050 (68). The underlying premise of the exhibition, that science and technology will provide solutions to climate change and its disruptions, carries the normative assumptions found in other types of trans-humanism. It does not engage climate as a complex socio-bio-geo mix of factors; rather it continues the trope of categorising nature as a fixed object to be controlled by reason and innovation. Cameron (2017b) identifies a number of normative messages in *Atmospherics* including:

> Carbon or CO2 is the climate question and reducing our carbon footprint is the sole and most effective response and course of action.
> Capitalist economics and economic growth can continue in a business-as-usual fashion if alternative energy and cleaner sources are made available.
> The atmosphere can be controlled if put back in balance.
> (70)

Cameron's alternative way of approaching climate would include a range of actants rendered invisible in the Science Museum exhibit – actants that together form entangled networks of human and non-human relations or natureculture. Climate cannot be usefully understood as a linear cause and effect event remedied by returning to a stable atmosphere. The list of forces Cameron (2017b) cites as actants composes a biological, geological, social network.

> Water; clouds; nitrogen; oxygen; carbon dioxide; coal; lifestyles; legislation; radiation; heat; gravity; profit motives; electrons; electricity; climate science; computer modelling; economic theory; carbon taxes and emissions trading; pollution; coal fired power stations; trees; energy policy; ideologies; rationalities; technologies; beliefs; thoughts; desires; consumption; small island states; malaria; cyclones; oceans; ice; United Nations Framework on Climate Change and nation states.
>
> (71)

Museum initiatives

A 2020 issue of the journal *Museum Management and Curatorship* reviews and surveys climate museum initiatives and online networks. Included in the discussion is Climate Museum UK (CMUK; climatemuseumuk.org), a digital and mobile activist museum with the explicit mission of expanding knowledge of the earth crisis: 'Naming the organisation [in May 2018] was a declaration of emergency, a call for support and a pledge of service in response to the urgency of the Earth crisis' (McKenzie 2020, 672).

The initiative not only challenges the neutrality of the museum but also confronts the orthodoxy attached to the role of the museum as a collection of physical artefacts, a legacy not to be underestimated, 'for the most part the value of [a museum's] material assets as their most unequivocal distinguishing characteristic is seldom challenged' (Shelton 2013, 483). Tom Jeffery (2021) argues for a fundamental ontological change to disrupt the untouchable position of the museum collection; the current ICOM principles of acquisition he says provide a 'shield to the deep roots of museological dualism in neoliberalism' (61).

The projects undertaken by CMUK and its associates present alternative relations and settings for engagements with non-human and inorganic actants. These undertakings reflect a new common sense or knowledge. Playful activities are advocacy for learning toward radical action. One project led by a CMUK associate based at Phytology Bethnal Green Nature Reserve asks: If you could talk to plants and trees what would you ask them? Starting conversations with plants is a way of 'considering the limitation of the human experience, language, and our struggle to avoid being human-centric'. Another CMUK project *The Wild Museum* invited participants at a Timber Festival in the National Forest in central England to hear stories about the forest's history of coal mining and landfill while making terracotta seeds, fungi, micro-organisms, worms and insects.

The projects of CMUK as a distributed experimental museum are clearly framed by the challenge of tackling 'climate and ecological emergencies together' (climate-museum-uk-newsletter-solstice). They advocate for activism including the civil resistance organised by *Just Stop Oil* (https://just stopoil.org/) with the October 2022 protest Occupy Westminster.

In 2011 American artist Amy Balkin initiated an archive of donated objects *A People's Archive of Sinking and Melting*. The widely disparate and otherwise ordinary objects are 'contributed by people living in places that may disappear because of the combined physical, political, and economic impacts of climate change, primarily sea level rise, erosion, desertification, and glacial melting' (Haigney 2021, 4). They include small items delivered to Balkin with a note about its relationship to the contributor in the context of a sinking or melting place. In a review in *The New Yorker*, the project is said to raise the challenge of archiving the climate crises. In a sweeping array of ephemera and personal items it seems to grasp the complexity of a global change, 'which touches everything, which seems to be unfolding both rapidly and slowly – through sudden disasters and also on the scale of geological time – drawing any kinds of boundaries feels impossible' (5). Balkin touches on the potency of the Archive as 'standing in as proxies for the contributors' recognition of the geopolitical production of precarity and slow-onset dispossession. Together, the contributions form one material record among many, a collection of community-gathered evidence, a public record, and a midden' (Hannah and Krajewski 2015, 41).

Attached to fragments of red stone, one contributor note reads:

> This firebrick was collected near the site of a house on Constance Estate, Trinidad and Tobago, that was destroyed and claimed by the sea. The ruins are still visible in the waters of Columbus Bay.
>
> (Haigney 2021, 5)

There are objects from all over, including Antarctica, Greenland, Cuba, Italy, Russia and Tuvalu. There are also objects with a relationship to Hurricane Sandy, which hit New York in 2012.

Some place

It was in the aftermath of Hurricane Sandy that the first climate-dedicated museum in the United States, the New York Climate Museum, was established in 2015. Its director Miranda Massie, a civil rights lawyer, was 'astounded a climate museum wasn't already underway' (Huntley 2020, 25). From the few exhibits the museum installed around the city, Massie observed that people who visit want an experience that offers 'guidance on how to act, especially in collaboration with other people' (25). They don't want more science. As Rebecca Huntley argues, the general population is ready to move into a climate-active stance if we can provide them with spaces and opportunities to do so with a sense of social safety (27).

In this respect, citizens are prepared for museums to 'stay with the trouble', to borrow Donna Haraway's compelling phrase. Haraway's once wayward ideas are now acknowledged as responsible and rigorous, such as her call to develop odd kin with non-humans:

> … we require each other in unexpected collaborations and combinations, in hot compost piles. We become-with each other or not at all. That kind of material semiotics is always situated, someplace and not noplace, entangled and worldly.
>
> (2016, 4)

Many museums already gesture toward such kin-making, but this invariably feels too disciplinary specific and neat. Odd kin, as Haraway conveys, are a messy composition of ideas, materials, feelings, artists, scientists, researchers, policies and communities to challenge the Capitaloscene with those politicians, billionaires and shouty media who peddle the new culture of greenwashing to support their own interests in fossil capital.

In the project of communicating change, museums must loudly detach from fossil fuel industries. A move so clearly necessary we must stop and wonder why, despite continuous protest, museums do not do so. Museum boards and management claim it is about funding, which Janes and Sandell (2019) reject arguing that perceived shortages have become an excuse for maintaining the status quo of the museum-as-mall: 'One toxic expression of this material fixation is the incessant talk of shortage in the museum world – be it money, staff, technology, or public support' (2).

A history of artist activism and intervention protests the sponsorship of museums by corporations that seek to 'art or science wash' ecologically unsustainable mining activities and other operations. Artist Hans Haacke pioneered institutional critique including with his didactic banner work *MetroMobiltan* (1985) that drew attention to the dubious relationship between Mobil Oil (now Exxon Mobil) and the Metropolitan Museum of Art, New York (Chong 2012, 105). The banners highlight 'the interrelationship between violence in Apartheid South Africa and Mobil's continuing supply of oil to the South African military and police' (Robins 2013, 96). Art Not Oil is a more recent collective of artists,

> designed in part to paint a truer portrait of an oil company than the caring image manufactured by events such as the BP Portrait Award [at the National Portrait Gallery] … and other such "cultural activities" which also happen to divert public attention away from their actual activities.
>
> (Chong 2012, 106)

In stubbornly continuing to accept fossil fuel sponsorship it is not viable for museums to support the change of mindset pursuant of new knowledge.

Climate Museum UK (climatemuseumuk.org) acknowledge this by carrying the logo *Oil Sponsorship Free*. Derreck Chong noted in 2012 that the BP/Tate's two-decade long partnership was not viable, as following the 2010 Gulf of Mexico catastrophe of the Deepwater Horizon drilling rig. 'BP is cast as not being socially responsible, with Tate drawn into a carbon-based economy that lends weight to Big Oil' (Chong 2012, 104).

Water will be here

From their analysis of Anthropocene exhibitions, Isager et al. (2021) observe that it is rare for exhibits to explicitly confront what human responsibility and transformation consists of. Two art exhibitions that do so, however, are *Placing the Golden Spike Landscapes of the Anthropocene* (2015 Institute of Visual Art, Milwaukee), which extends into an urban greening project, and *Let's Talk about the Weather* (Sursock Museum, Beirut, 2016 and Guangdong Times Museum, 2018).

The former exhibition investigates when and where the Anthropocene began, finding this 'Golden Spike' exceeds any precise moment or place and is distributed. It can be found in the material residues of digital spaces and in domains construed as immaterial including 'in the anxious minds of the alleged agents of the Anthropocene' (Hannah and Krajewski 2015, 23). In the exhibit, the artists and contributors do not shy away from the element of fear and the everyday. Eric Corriel's work *Water Will be Here* (2010) exemplifies this approach. In his installations, each night as the sun sets, a wall of water is projected against the museum windows in front of which people assume postures that give a terrifying normality to the extreme events.

The museum is well positioned to link issues and convey the entanglement of subjects that used to be treated as distinct. Historian Dipesh Chakrabarty asserted in 2009 that human history is no longer separate from geological time. He more recently sees this as 'an emerging role for museums: to create a new story of humans in which our "natural" and "civilisational" or "technological" histories are blended' (2019, 18). In Chicago where he lives, he finds it no longer appropriate that museums (The Field Museum of Natural History and the Museum of Science and Technology) remain apart physically and conceptually. The deep and fundamental separation of these museums on a conceptual level must break down (18).

The entanglement of once non-aligned types of knowledge that he gestures toward is a thriving investigation among artists, researchers, critical theorists and others who increasingly discern the nuanced flatness of ontological relations between things. This is a widening awareness with many stances having arisen through the humanities developing links between climate change and globalisation: 'A fundamental shift has happened in our very awareness of the position of humans in the order of things on this planet' (Chakrabarty 2019, 15).

The hopeful opportunity for museums is that they are not unfamiliar with change brought about by necessity; during the 1980s the museums' institutional relevance was challenged by the new museology. The confrontation with colonial, gendered and class-based legacies and practices was a 'major revolution' that found museums 'define themselves as not just repositories of objects or even memories, but as sites of particular experiences for visitors' (Chakrabarty 2019, 14). The many case studies that emerged raised a mirror to the museum through an institutional critique that demanded the inclusion of previously misrepresented, overlooked or ignored peoples and their histories.

The current moment again requires deep provocation and there are philosophical sources that can be garnered to support transformation. Some sources come across as abstracted from the 'real' world yet there is an overarching realisation that thought itself has to transform. It is the case that philosophical discourse can seem lacking in practical, real-world applicability. Tim Ingold (2021) muses:

> I have always been slightly bemused by scholars who bury their heads in the most arcane and impenetrable of texts in the effort, they tell us, to get to the bottom of our experience as beings in a world. You would think the best way to fathom the depths of human experience would be to attend to the world itself, and to learn directly from what it has to tell us.
>
> (8)

Enter the museum. Museums that provide experiential encounters with the material knowledge developed through philosophical debate and discourse will be vital participants in the responsible transformation of obsolete humanist thinking. As Cary Wolfe (2010) says, it 'will take all hands-on deck … to fully comprehend what amounts to a new reality: that the human occupies a new place in the universe, a universe now populated by … nonhuman subjects' (47).

Beyond innovation

In comprehending a new reality, museums can usefully move beyond the hype attached to innovation, and value the less fashionable mode of maintenance. Sociologist Ulrich Beck characterises the new globalised era of 'reflexive modernity' as a world that is left with 'zombie institutions, practices and thought constructs that no longer respond to reality' (Boulton 2016, 775). Many universities, for example, have incrementally moved away from their significance as critical response that speaks truth to power in order to secure research funding from industry through the promise of innovation. Academic research is subsumed into conformity by an excessive managerialism (Joseph 2015) that supports economic growth pitched through a rhetoric of innovation.

The rise of populist anti-intellectualism has significantly impacted on the academy including the closure of humanities courses and a corresponding diminishment of critical thinking. Contributing to the demise of the academy in advancing critical debate and dialogue is the commodification of online learning in higher education institutions, and short-term cost-cutting measures such as deaccessioning physical library collections. The university has committed to a transhumanist path that aligns with the neoliberalisation of culture and society. Museums should carefully evaluate this development and provide an alternative vision to transhumanist neoliberal innovation-ism.

Technology historians Andrew Russell and Lee Vinsel (2016) argue that overvaluing innovation has become detrimental to the social good. They trace its prominence as a post-1960s discourse generated to replace the moral complexities attached to progress. Innovation was a useful notion as it was viewed as morally neutral: 'it provided a way to celebrate the accomplishments of a high-tech age without expecting too much from them in the way of moral and social improvement' (3). The strategic turn of institutions away from critical discourse and toward innovation is linked to reliance upon, rather than caution toward, the products of Big Tech. Russell and Vinsel argue the need to move scholarship away from innovation and toward the social function of maintenance: 'A focus on maintenance provides opportunities to ask questions about what we really want out of technologies. What do we really care about? What kind of society do we want to live in? Will this help us get there?' (15).

A transforming counter move for museums is to expose how a mantra of technological innovation masks the inattention given to maintaining physical systems. Moving beyond the innovation mantra is a productive and relevant stance for museums given the hyperactivity around online platforms. For museums to evaluate rather than conform to ubiquitous screen culture and digital dependence and to focus on embodied encounters with the material world will signal that the loss of analogue encounters is detrimental. The analogue remains the reality of everyday relations with the material earth and more-than-human others.

Media philosophers Robert Hassan and Thomas Sutherland (2017) are concerned at the non-equivalence in digital computing between humans and nature. They write that digital logic is '"unnatural" in the most literal sense in that its logic moves us towards a virtual world that has no analogue in the complex ecologies of organisms that comprise life on Earth – and of which humans are a component part' (2017, 10). Here is where museums should be wary of documenting climate change as a digital spectacle. It is improper for crises and tragedies to be commodified through the use of digital media. Unfortunately, this is the case at The Museu do Amajha/Museum of Tomorrow in Rio de Janeiro, a science museum focused on the environment – a dubious spectacle in a country where activists for environmental change are regularly murdered (http://blogs.edf.org/climatetalks/2011/06/02/brazil-at-the-crossroads/).

The Museum of Tomorrow does not engage with the deforestation of the Amazon or with the massive pollution in the bay where it is located nor with Brazil's tragic history of mining disasters. A video of mining waste spilling from a ruptured Samarco tailings dam in Mariana was shown at the museum, but as this spill is considered the biggest environmental disaster in Brazil's history, this is clearly an inadequate response. The 2015 Samarco dam disaster has been classified as a violation of human rights. BHP Billiton and Vale own Samarco as a joint venture. The conflict of interest between Vale and the Brazilian government has biased investigation of the dam, which was known could collapse.

3 Repurposing the inclusive museum

The observation is that the 2020s are 'on track to become the era of museum activism' (Message and Foster 2019, 617). Institutional critique will increasingly focus on activism and crises exemplified by Janes and Sandell's edited collection *Museum Activism* (2019). With the unfolding of climate action, the museum that retains a stance of neutrality will be increasingly irrelevant. Climate activism as responsible governance for museums aligns with inclusion of our assemblage with the inhuman. It is not sufficient to engage with the climate emergency while remaining bound to representations of exceptionalism that overlook human entanglements in the material realm.

One of the areas that is challenged by the climate emergency is the model of cultural inclusion that has been prevalent since the 1980s. Inclusion is filtered through the cultural politics of identity, limiting the entrance of more-than-human actors, assemblages and encounters. Acknowledging that cultural inclusion in museums has become an exclusionary gesture does not mean that inclusion agendas should be abandoned. It does however highlight that difference, when channelled into the politics of identity, should not obfuscate a deeper collective entanglement with earth systems. A united and cross-disciplinary attention to the material world responds to communities increasingly dispossessed by climatic events – refugees, stateless people, animals, soils – and communicates the implication of ever more widening gaps between rich and poor. As Fredengren and Åsberg (2020) write, the injustices of identity 'need to meet up with the bias of human supremacism in times of great injustice to vast ecologies of nonhumans' (65).

'Inclusion' like 'connectivity' has become aligned ideologically and diminished through appropriation by neoliberal forces. The just intention to facilitate greater equity through inclusion is an idea that has been commodified, requiring a new inclusion paradigm. In such a shift, the museum moves away from the instrumental value attached to inclusion and in doing so enables practices that depart from usual engagements with non-human agency. Through the act of seeking a contrary path, institutions expose the populist rhetoric of Big Tech, Big Mining and the advocates, lobbyists and think tanks supporting powerful global elites. Connectivism sounds like a

DOI: 10.4324/9780367741945-4

form of inclusion but is better approached as 'An organizational logic, it is the promiscuous inclusion of seemingly unrelated elements' that among other appropriations supports Big Tech; it 'drives Google's geopolitical strategy of global influence, which proceeds through a techno-affirmative desire to annex everything' (Culp 2016, 67). Here we return to the ubiquitous discourse of trans-humanism. As Andrew Culp argues this techno-affirmative world needs to be replaced by authentic engagements with difference.

The idea of the universal museum can be reframed to support a new status for inclusion. Traditionally the purpose of the universal museum is not ontological. Rather it is to be a dynamic space 'formed through contact and exchange with diverse people' (Cuno 2011, 3). James Cuno writes that 'encyclopaedic museums bear witness to the truth that culture is hybrid and mongrelized, evidence to the intertwined history of cultures and the connectedness that has always marked our globalized history' (85). In this fair praise what is valued is concern for a shared human endeavour and progress. There is no proper sense of the other-than-human. The type of universal museum of enlightenment that Cuno advances does not confront the steady and ongoing damage to life on earth. It does not confront the difficulty that Cary Wolfe (2010) poses, that 'Enlightenment rationality is not, as it were, rational enough, because it stops short of applying its own protocols and commitments to itself' (xx).

The meaning behind the universal museum is deeply tied to the ideal of a 'shared humankind as heir to a single shared world heritage' (Dibley 2017, 36). Ben Dibley contests the approach as it requires a cosmopolitan citizen fitting into an already existing common world, when there is no such world. Yet it is this idealised shared world that is invoked in the museum statements he identifies, particularly the Declaration on the Importance and Value of Universal Museums (ICOM 2006) and the Buffon Declaration (2007). Dibley's evaluation of these museum manifestos observes they have a too narrow understanding of human relations with the non-human. He also brings to the discussion research showing that citizens may declare empathy for people suffering from climate disruption, but this is a vacuous empathy that does not lead to action (43). The gap between empathy and action is one that is not even hidden by respondents who acknowledge their compassion is passive. What is highlighted is that a universal shared heritage is an abstract idea that 'offers no necessary impetus toward a cosmopolitan momentum'. Moreover, 'it is as likely to lead to parochial retreat as it is to cosmopolitan openness' (44).

The commodification of inclusion culture

The museum 'turn' to inclusion, particularly by encouraging visitor participation, has become a movement. 'Audience participation is *de rigeur* in museums today and in society more generally' (Tøndborg 2013, 10). In museums, participatory practice has been formatively influenced by Nina Simon, whose 'go-to-guide *The Participatory Museum* (2010) and complementary blog Museum 2.0 have started a virtual participation movement' (Tøndborg 2013, 10).

The participation-as-inclusion movement in museums responds to a wider development influenced, at least in part, by Jacques Rancièr's notion of an 'emancipated spectator' and Nicholas Bourriaud's promotion of relational aesthetics (Tøndborg 2013). In tandem is the view that an inclusive curator is an impartial facilitator, rather than an expert who defines and articulates her own position in relation to an issue. Lynn Wray (2019) investigates how a trend toward curatorial anti-authorialism in art museums has arisen from adopting approaches of Chantal Mouffe and Slavoj Žižek as well as Rancière.

Tøndborg and Wray offer critical perspectives on the potential and limits of a 'post political' turn that is impacting the work the museum performs. Wray (2019) notes that one difficulty with anti-authorial approaches is that 'by positioning the curator as a neutral facilitator, they actively reinforce the idea that museums are and ought to be apolitical' (319). An approach that supports the inclination of museums to appear neutral on climate change. Examining the efficacy of inclusive approaches, Tøndborg (2013) concludes that 'museums do not appear more democratic and inclusive just by replacing knowledge, or aesthetic knowledge, with dialogue and debate' (13).

A thin line is discernible between inclusion, populism and controversy. The Malmo Museum and Skelleftea Museum in Scandinavia pioneered what they brand as 'Hot Topic' exhibitions, (on rape, obesity, drug addiction and animal rights, among others). The hot topic exhibition format exemplifies a shift that replaces learning with awareness-raising that, arguably, 'turns the museum into a soapbox' (Tøndborg 2013, 7). Suggesting the need for further critique of this trend, which is intended to initiate dialogue, Tøndborg conveys the difference between a visitor for whom 'all that is required ... is to participate', a requirement that particularly suits visitors 'accustomed to taking part and sharing their thoughts through social media' (11) and genuine participation. The latter is a more active involvement and reflects Claire Bishop's 'invention of an "unpredictable subject" who momentarily occupies the street, the factory, or the museum – rather than a fixed space of allocated participation whose counter-power is dependent on the dominant order' (cited in Tøndborg 2013, 15).

Wray (2019) argues that inclusive exhibits that hope to generate multiple visitor voices as a method to undo or avoid the hegemonic power of the curator are a problematic means of activating audiences. She identifies three particular problems:

> Firstly, anti-authorial approaches often create new barriers to comprehension that can work to alienate, rather than empower, exhibition visitors. Secondly, they fail to work with the unique attributes and limitations of the exhibition medium that make them such potentially mobilising forces in the first place. And finally, anti-authorial approaches

reinforce the impression that the museum is impartial and objective. This can only work to inhibit the process of critical consciousness raising that the curators are striving for.

(318)

Illusionary participation

An implication of the participation-inclusion trend is that it has become the hegemonic frame with which to ascertain whether a museum practice is valid. As Wray suggests it does little to dispel the myth that the museum is a neutral institution. For Robert Janes (2015), the 'magical belief of neutrality' (2) is 'a pronounced liability' for museums when it comes to being effective on climate change and its disruptions. While, as we have seen, there are some museums prioritising the climate emergency and seeking to differently interpret human and climate relations, this is not the norm, 'vanity architecture projects continue, the yearning for popularity remains obsessive, and the plutocrats (a significant portion of whom are climate change deniers) continue to play in the world's museums' (Janes 2015, 3).

The neoliberal appropriation of participation-inclusion rhetoric commodifies inclusion, to the detriment of new inventions of what it means to be human in response to climate. The well-intended agenda of inclusion sought wider community and minoritarian access to museums; however, the effect of this initiative was often illusionary. Bernadette Lynch (2010) observes that 'critical pedagogy' in expanded museum education departments that were part of the participatory turn of the new museology, 'too often provided an illusion of participation' (9). Alongside the expansion was the rise of neoliberal public policy based on deregulation that saw 'inclusion', particularly in museum education and outreach initiatives, tied to economic markers of success.

What began in the new museology as a commitment to critical pedagogy evolved into the economic use of 'inclusion'. Formulated as Key Performance Indicators, measures of 'inclusion', 'diversity' and 'participation' were suddenly everywhere, influencing exhibition programs, staffing and community outreach. During the 1980s and 1990s, complex structural inequities were translated into inclusion rhetoric via museum strategic plans commissioned from the Big Four accounting firms, with mission statements and glossy education packs produced by public relations consultants, and sponsorship and marketing teams.

Measuring inclusion is a gesture to the way that difference is politicised, particularly when it slips into the politics of cultural identity. Instead of exposing inequity, there is a narrow engagement with inclusion through the concept of empowering the individual. The status quo is accommodated through representations of the self and other that do little to expose the effect of the divisive notion of 'us' and 'them'. Participation-inclusion that rhetorically acknowledges and assimilates the 'other' does not change the status quo or support collaborations that resist the inequitable structures and legacies of that status quo.

The museum can ask whether 'the universalising of inclusion escapes the subordination of cultural difference to the dominant culture or unwittingly repeats a narrow identification of identity albeit through a shifted ideological compass' (Baker 2017b, 3). An inclusive museum that is able to respond effectively to the enormity of the disruptions caused by changing climate has to be one step ahead of the commodification of an idea like inclusion. And be wary of too easy categories that champion individual well-being and diminish the possibility of achieving real structural change. Attention to the complex alterity of the non-human as part of 'us' (the self in the other) is imperative to foster this relation. Whomever 'we' are, we are coextensive with the other. It is not about making the 'other' the same as us, but rather about our difference from the inorganic worlds of which we are a part. This is the defamiliarisation event that has to occur.

Cultural inclusion as neoliberal hegemony

In her evaluation of museums that are unhelpful in response to people categorised as 'marginalised' and 'vulnerable', Lynch (2010) says there are lessons to be learned for museums in relationships with all people. I suggest the lesson extend to the marginalisation of more-than-human assemblages', be this soil, elements or the weather. With a more generous awareness of inclusion, previously devalued agents become observable and in doing so neglected justice and equity issues are harder to ignore. While this direction raises the fraught area of what to prioritise in an assemblage, the acknowledgement concedes the need for a degree of unrecognisability as forms of agency come into proximity that have been kept apart. Useful is Tim Ingold's discussion of the term 'complicate', from the Latin *com*, 'together' plus *plicare*, 'to fold' (Ingold 2021, 20). Ingold (2021) writes how material things are always in processes of folding together; they are always complicated and crumpled, whether a boulder or a tree. Humans can't tolerate such a disorderly world, he says, because

> crumples are alien to our desire for order. We prefer a world that answers to the call of reason. Whatever we build or make, we endeavour to straighten things out, to simplify. We like external surfaces to be smooth and flat, and angles sharp.
>
> (20)

The useful museum will increasingly be a crumpler.

Lynch (2010) observes the importance of heeding pitfalls that can arise with genuine attempts at fostering a collaborative relationship with those deemed 'marginalised'. The radical change she wants to foster in museums is a 'profound sense of interdependence' and solidarity' (15). This is neither 'about kindness or charity toward an "other", but, rather, recognising the self in an Other' (15). I propose garnering this profundity to readdress engagements with geo 'othered' existents, to fold thought of the self into the inorganic realm.

What is masked by a rhetoric of empathy toward the marginalised is actually deep-seated injustice: 'One is repeatedly struck by how it is that museums continue to define the rules of engagement, avoiding conflict and the active agency of participants to express it, while claiming open "negotiation"' (Lynch 2010, 6). The aim of a useful museum, Lynch argues, is the 'radical shift' (7) of the 'marginalised' becoming agents of change. This involves collectively taking ownership of museums to overcome barriers to transformation.

To reach this milestone is to understand that inequities, rather than real change, are often inadvertently perpetuated. A situation that is not dissimilar to hubristic Anthropocene discourse in the exhibits discussed in the previous chapter. There is a significant gap between proper change that involves unfamiliar and difficult thinking, and more comfortable adjustments that are an illusion of change.

At the movies

In evaluating the commodification of inclusion in museums, I elsewhere liken affecting museum encounters that are transformative to the appeal of the cinema (Baker 2017b). The 'cinematic museum' is a concept that confronts the inadequacy of institutional critique that overlooks the capacity of museums to provide unexpected relations with difference.

In movies, potent museum experiences frequently arise through strange, weird and dangerous lures that make audiences squirm – eyeballs in jars, creepy portraits, monsters in basements. In the movies, museums and galleries are counterintuitively not places to encounter order and rationality, but instead present strange yet desirable forms of alterity. More-than-human affects that threaten a character's logical reasoning stimulate a desirable abjection (Janice Baker 2013). The once whole and stable self transforms into something considerably less certain in cinematic encounters. The wayward antics of artefacts, specimens and collections are transforming events that facilitate a change in human relationships with the object world. The critical insight, that 'we' are not entirely 'human', arising from such imaginative entanglements is overlooked in the museological literature where the focus is more frequently on measurable analytical and ideological framings of the pedagogical museum experience.

An expanded meaning of inclusion is one that encompasses the entanglement of one thing as another in ways that give a kind of proto-consciousness to matter, ecologies and climate. It is not about assimilating difference so that which is different is made familiar through being the same as us. Museums can communicate the quantum leap in thought that this involves in the sense that theoretical physicist Karen Barad (2017) articulates:

Repurposing the inclusive museum 57

> Entanglements are not the mere intertwinings of, or linkages between, separate events or entities or simply forms of interdependence that point to the interconnectedness of all being as one. Entanglements are the ontological inseparability of intra-acting agencies...Phenomena are specific material relations of the ongoing differentiating of the world, where 'material' needs to be understood as iteratively constituted through force relations. Phenomena are not located in space and time; rather phenomena are material entanglements enfolded and threaded through the spacetimemattering of the universe.
>
> (111)

The type of inclusion that extends the hyper individualism of the modern human does not support the generative potential of this anti-humanist stance. A stance that enables a move from the Cartesian notion 'I think therefore I am' to the anti-substance metaphysics that Karen Barad conceives as spacetimemattering.

The privilege of thinking that life is an autonomous entity that exists over and above inorganic matter reflects the continuing influence of Kant's judgement that nature only exists because humans conceive it as such. Kantian judgement is a challenge for the new inclusion, as it delimits the ability to think life:

> It goes without saying that the major challenge that faces philosophical reflection on the concept of life is not about the most accurate or coherent definition of life. Life as a concept quite effortlessly passes between the poles of reductionism and mysticism – life can be defined down to the molecular level, at the same time that the notion of the irreducibility and mystery of life raises the concept up to the existential and spiritual levels. Instead, the major challenge for any ontology of life lies in being able to think its very conditions of being thought at all.
>
> (Thacker 2010, 250)

To think what thinking is takes many of us into the realm of a radical alteration of who we think we are. But this is a necessary alterity. Critiques of the modern human's excessive self-confidence have initiated a geologic turn that acknowledges that 'material realities of life on the planet are outpacing our ways of knowing' (Ellsworth and Kruse 2013, 14).

Radical alterity has many fronts. In spacetimemattering Karen Barad, for example, understands 'land' not as property or territory but as time-being with its own memory (cited in Fritsch et al., 2018, 15). Ellsworth and Kruse (2013) observe that as categorical distinctions break down between geology's external works (rocks, minerals, mountains) and life, 'it's possible to claim, without taking too much poetic license that we humans are walking rocks' (23).

In arguing that 'urgent action is needed to produce knowledge and cognitive frames that will give rise to new ways of thinking' in relation to climate change, Fiona Cameron (2017a) exposes the modern worldview based on Cartesian rationality as increasingly problematic (16). The ethics of care for the non-human world cannot shift without a mindset that is able to break with modern human-centred views on climate and the environment. The move away from a universal humanist museum toward ontological realms explored by relational thinking enables 'new ways to represent, talk about humans, non-humans, culture and cultural diversity, heritage objects, the environment and climate change as new types of narratives, sets of practices, and concepts' (Cameron 2017a, 16).

Hence, a need for the parameters of inclusion to move outside the 'us and them' opposition that recurs in institutional rhetoric. Oppositional thinking arises when the objective of an exhibit is neatly represented as the other, rather than the object voicing its agency. Lynch experiences this in museums that perpetuate the status of the 'marginalised' rather than as collaborators and agents of change. Similarly, it is too easy to 'other' the climate as distant information out there, rather than as local, visceral and ongoing mixes of intra-acting phenomena.

Climate inclusion and popularism

Anchorage Museum director Julie Decker (2020) addresses a concerning popularisation attached to narratives of global warming in relation to a new colonisation of the Arctic periphery (640). Climate impacts are romanticised making the Arctic trendy and a venue for celebrity events such as Oprah Winfrey's travel to Alaska, 'posting Instagram photos with microbrews' (640). Museums are too easily complicit in this trend with

> many participating in a race to have an exhibition about climate change and the Arctic, from the British Museum to the Museum of Craft and Design in Los Angeles to the Brooklyn Museum in a span of just one year in 2019-2020.
>
> (640)

Arctic residents of Alaska are relocating their communities as shores and islands sink and disappear with the thawing permafrost. Decker cautions against a proliferation of urban perspectives of the melting North, a tragedy that is real and present. A cross-institutional initiative led by architects and urbanists called *Countryside, the Future* at New York's Guggenheim Museum is the type of city-based initiative that Decker is wary of.

> While intriguing and not necessarily invalid, it is, from a broad view, the continuation of a colonial trend. Outsiders from urban centres look out to the periphery and draw conclusions for export: outsiders as voyeurs, experts of places not visited or lived, and knowledge extracted. The outsider view as the authoritative voice over the insider view is centuries old

in Indigenous places, such as the Arctic. The Arctic vernacular is not perceived as holding the knowledge; instead, the knowledge is somehow applied from the outside.

(Decker 2020, 641)

The cautionary message here communicates the difficulty with a too superficial gesture toward inclusion; the multiple intra-acting agents that comprise the Arctic are not included in the story of climate change and the point of view does not carry felt cultural and environmental loss. A loss better acknowledged through a wisdom of powerful myths, fables and oral histories that enact local visceral entanglement with the melting earth.

Robert Macfarlane (2020) writes with the power of such myth never far from his observations of the melting permafrost in Greenland, a place also experiencing unweather, that is, weather so extreme it seems to come from somewhere else (334). In the village of Kulusuk, Greenlandic life has significantly altered 'the inhabitants of this village are part of the precariat of a volatile, fast-warping planet' (335). With the loss of intricate cycles of ice, traditional life across the Arctic is under threat of erasure. Macfarlane says the word being used by the Inuktitut of Baffin Island in the Canadian Arctic is *uggianaqtuq* or to behave strangely. The word 'refers at once to the changes in the weather, the changes in the ice, and the consequent changes in the people themselves' (335).

The melting cryosphere in Greenland is revealing haunted places as long buried things come to the surface. In the hot summer of 2016, 'Reindeer corpses that had died of anthrax seventy years earlier were exposed to the air. Twenty-three people were infected, their skin blackened with lesions. One a child died'. While in north-west Greenland, 'a buried Cold War US military base and the toxic waste it contained began to rise, troublesome history thought long since entombed is emerging again'. When the base was abandoned in 1967 the infrastructure was left beneath the ice, 'including the biological, chemical and radioactive waste it contained' (Macfarlane 2020, 330).

Activists and theorists who translate complex physical systems into compelling metaphors in relation to anthropogenic climate change are finding a wide readership. But there is a need to problematise rather than simplify what is not human. A shift to understand that planet earth is an ongoing process of differentiation, and as an agent of change, we should listen to what it has to say. An active listening that takes western perception into unfamiliar territory. On this, people are prepared for museums to provide unsettling experiences that mark their own increasing awareness of damaged environments and ecosystems.

Philosophers Graham Harman and Timothy Morton, advocates of the stance known as object-oriented ontology, aim to nuance the agency of objects in ways that support alternative approaches to presenting material culture. We have already encountered Morton framing climate change as a hyperobject – an object too large to be seen, nonlocal yet everywhere. A hyperobject, such as the warming globe, 'demands a *geophilosophy* that doesn't think simply in terms of human events and human significance' (Morton 2013, 7). Once considered a stance for philosophers to debate, perspectives such as object-oriented ontology are recognised today as imperative by science. 'The attempt to take account of the interests of objects – of the non-human – is essential if we are to move forward a sustainable energy ethics' (Murphy et al., 2021, 4).

The museum traditionally interprets nature through presentations of human events and significance. Even when not the intention, nature is staged to distance human life from the earth. In relation to 'natural' resources extracted for human use, there is acceptance that this is the natural way of things. A phobic blind spot when it comes to the material world; a narrow vision that can only see a stark distinction between human being and the geological (Baker 2017a). A failure of imagination.

The wondrous ordinary

Within a museum's corridors and galleries, a fold in expectation can deepen a sense of the mysterious thingness of objects. The curated space of the gallery cannot fix or force a response from a viewing public and despite careful curatorial planning is a place of multiple and partial connections continually pulling in all directions (Hetherington 1997, 214). These experiences don't have to be grand or spectacular, and in an era of screen wizardry and digital pomp are often to be found in the ordinary and banal.

An ordinary fold occurred during my visit to the geological museum attached to the Western Australian School of Mines in the gold mining town of Kalgoorlie-Boulder. Founded in 1907, the collections in the high-ceilinged room with its original wood and glass display cabinets are identified by hand and typewritten labels. Pleasingly undisturbed by anything digital, the hall is a step back in time, a museum of a geology museum. An anomaly among the displays are large rocks beneath the vitrines. The rocks are too heavy for the antique wooden cabinets so are pushed underneath where they are not part of the taxonomic arrangement of the rock and mineral specimens viewed through glass. They are a fold in the geological story and revealed to me the rather obvious yet overlooked fact that rock specimens are ordered displays that are nothing like their presence *as* the earth.

The escapee rocks tucked beneath the display cases function as in-between objects that are collected yet unclassified. There is something poignant about this. Perhaps it has to do with the way in which overlooked things drag us into no-where spaces of thought. They are alone and unorganised and this affects a kind of empathy. Anomalies like these rocks evade the order of things imposed on their neighbours. They pause the usual perception of things. Such unintended reverie is perhaps conducive with the idea of a gallery as a slow medium, a space offering a slower pace for reflection than is familiar in a rapidly changing world (Robin et al., 2014, 208).

In observing anomalies and the capacity to understand difference through objects, vigilance is needed to avoid falling into a privilege Manuel De Landa ([1997] 2019) calls 'organic chauvinism' – a habit or way of constructing meaning 'that leads us to underestimate the vitality of the processes of self-organisation in other spheres of reality' (Yusoff 2013, 790). Even though we might seek to escape the usual order of things, the escape is still perceived through habits of reason and language. Kathryn Yusoff (2013) speaks to this difficulty through her thinking on geologic life, 'a mineralogical dimension of human composition that remains currently undertheorised in social thought …' (779). We are organised with geology, she seems to say. Hence, yet again, we confront the difficulty of western thought to acknowledge natureculture, and in this instance the perception that 'we' are lithic entities. Our thought is elemental and the elemental is thought. The difficulty confronted is the dilemma acknowledged in Derridean deconstruction philosophy, that we can only render codes of difference unstable, and that it is not possible to reach a new kind of other life (Fritsch et al., 2018, 8). Yet as Michael Marder (2018) offers this does not mean we should not 'accept the challenge of this impossible possibility' (164).

An elemental 'us' does not sit comfortably with the way behavioural changes in response to physical systems are framed. As Yusoff (2013) writes, in moving away from burning fossil fuels, 'behaviours are curiously cleaved off from the subject to be operated on … without any acknowledgement of how collectively, and to different extents, life has been constituted with and in fossil fuels' (792). With these significant challenges to business-as-usual selfhood, my rather ordinary response to rocks as escapees from the order of things is either an ineffectual new age conceit, or, more hopefully, a shift toward acknowledging that rocks are self-organised and exist with their own reality. It is with such openness to the existential reality of geologic forces as agents of change that alterity – alternative existence – is brought forth or enabled in engagements with soils, minerals and those existents known as fossil fuels.

It is intriguing as well as challenging to locate in the inclusion of the geological, a frontline for a new natureculture entanglement, a line of flight that departs from the conventions and assumptions of western humanism. It puts 'what is human?' under the lens as a practice to locate a more nuanced ethics for current crises. The museum can stage these new/old stories, and in so doing construct a contemporaneity that brings together stone lore and science. Of course, museums do already look closely at artefacts and specimens through an historical lens, but the looking always seems to slip into the anthropocentric. And this bias is how we understand knowledge. What happens however when an object peers closely at *homo sapiens*? When a meteor comes to earth? When rain does not stop? The new inclusion rejects the hubris of organic chauvinism as the dominant frame of knowledge-making. The museum that once used to manifest this inequitable privilege now has a very different and urgent function.

Ruin studies

Contributing to the new materialism by resisting the usual interpretation of what makes a ruin 'heritage', scholars are directing attention to previously overlooked material remains. In art, critical archaeology and critical heritage, apprehending a ruin or decay aesthetic is no longer to expect materials to offer a stable and solid ground upon which to prop human affairs. As Caitlin DeSilvey (2017) investigates in her search to understand a generative impact attached to heritage that is 'beyond saving', the key comes back to identity and subjectivity. To be open to new alignments with objects and architectures beyond their instrumental value is to 'unsettle our own sense of a coherent and unified self' (DeSilvey 2017, 183). It is to 'recognize that our identities are made through processes of subversion and fraying as much as they are through process of consolidation and stabilization processes' (183).

The new materialism is a disruption to common sense as it turns our attention to objects and things that are generally not valued. Museums too can intentionally look for otherness in material encounters rather than focusing on things being the same as the human world. It becomes about communicating feelings that may not sit viscerally with the status quo as they move outside the comfort zone of conventional value systems.

DeSilvey (2017) refers to a state of post-preservation to argue for a new way of understanding decay. An understanding that is able to leave things as they are so that material change can be experienced as a value in itself. This is counter-intuitive from the perspective of traditional heritage which values what remains 'because they are being saved' (DeSilvey 2017, 13). As she remarks, the real value of old ways of understanding heritage is that 'Objects of heritage are preserved, most transparently, in order to stabilize memory in material form and to stabilize associated identity formation' (13). Heritage involves practices of accumulation and preservation, when it would sometimes be more responsible to find ways of valuing the material past that instead 'release the things we care about into other systems of significance' (17).

In her poignant studies of the felt loss to communities of English coastal landmarks, DeSilvey details the long struggle to preserve a 1792 lighthouse at Orford Ness on the Suffolk coast and a harbour at Mullion Cove in Cornwell. As heritage that is beyond saving with rising seas and changing weather patterns, she asks what if the focus shifts to a type of care she calls palliative curation. What if the life stories of these objects' constituent materials are told.

> The iron, brick, and concrete that make up the bulk of the structure have biographies of formation, extraction, and transformation that precede their assembly in the ostensibly coherent shape of the lighthouse, and

these constituent materials also have a future, which will play out after the structure loses its current form.

(DeSilvey 2017, 173)

Through the method of bringing together cross disciplinary expertise, creative and critical thinkers are finding ways of acknowledging the existence of the inanimate outside the parameters of the modern subject and its objects. What this complements is the value of acknowledging First Nations ways of knowing animate and inanimate relations and putting this knowledge at the centre of museum theory and practice.

Drawing on Haraway's call to be present and stay with the trouble, and Povinelli's geontological framing as more relevant today than Foucauldian biopolitics, Alexandra Deem (2019) considers the parameters of real social transformation through First Nations experience. The stakes have changed she argues, and Foucault's epistemic breaks in knowledge are better framed by ontology. She uses the 2016 No Dakota Access Pipeline (#NoDAPL) movement to ascertain if online protest functions as resistance by creating a space that aligns with traditional ontology, in this case, that of the Standing Rock Sioux tribe.

Dakota and Lakota historian LaDonna Brave Bull Allard, who was the founder of the Standing Rock camp, showed how water threatened by the pipeline plays at the limits of Western notions of life and nonlife. The river is integral to the world of the Sioux, a relation that is ontologically incompatible to the U.S. government. As Deem (2019) observes, the Sioux

> imagine the livelihood of the river beyond its capacity to sustain human life in biological terms … Crossing the biological, cultural and natural domains, Allard's account subverts the Western systems of valuation based on (biological) life and death, as well as life and nonlife.
>
> (126)

For Deem, '#NoDAPL is a site of potential, an instance of digital decoloniality that, in the form of a quasi-event, subverts the possibility of maintaining difference in its current terms' (126). There is a powerful dynamic to notice here between unlike groups of activists that gestures to geo-inclusive spaces for museums to acknowledge and generate.

4 Museums, climate fiction and the anthropocene

The story telling in recent climate fiction illuminates stances on climate and supports an active role for museums in the governance of climate change. As climate change is a global emergency, all recent fiction in one way or another becomes about climate. If a work of fiction ignores or omits the warming planet and its significant disruptions, the text creates a parallel universe to what is happening in reality. The omission raises its relevance for creative practice, cultural forms and institutions: 'There seems to be a growing sense that architectures, infrastructures and artworks will risk being irrelevant at best, dangerous at worse, if their designers and makers don't take this newly salient reality into account and recalibrate their work' (Ellsworth and Kruse 2013, 20).

In Kim Stanley Robinson's prescient 2012 science fiction novel *2312*, the era of global refusal to deal with climate change is referred to as 'the dithering'. The dithering of course is the twenty-first century. Robinson's 2020 novel *The Ministry for the Future* is science fiction but located in an almost present future. The story opens with a catastrophic heatwave in India that kills 20 million people: 'And then the sun cracked the eastern horizon. It blazed like an atomic bomb, which of course it was' (2020, 2). The novel imagines a civilisation where a standing committee of the Paris Agreement comes together to work collaboratively to achieve a balance with the biosphere. Robinson argues that plausible bridges can be made now to prevent mass extinctions and a pathway of hope is generated in his novel that is absent in much postapocalyptic fiction.

Environmental philosopher Ted Toadvine (2018) values stories of ecological disaster that are situated in the present as a way of responsibly thinking about the impact of climate change (53). He considers that in fiction, as in life, there is a failure to live in the present, a 'temporal suspension of the present' (53). Disaster stories tend to leap between the past and the future, with the present a strange omission; hence the 'world here *as* sense' needs to be rediscovered (55). What we feel as we live each minute is here and now, not a past or future. To reach the embodied present, storytelling should be a realistic engagement with the earth, with seas, soils and $CO2$. The new realism that this present-ness enables will not resemble the fictional worlds

DOI: 10.4324/9780367741945-5

constructed by contemporary literature that acts as if it was written before climate change. Perhaps we can understand the new realism as a radical inclusion of the present.

Author and critic Amitav Ghosh calls an avoidance of climate in literary fiction *The Great Derangement* (2016). Ghosh considers that the lack of attention to climate change in literary fiction, literary journals and book reviews is a reflection of a broader imaginative and cultural failure that lies at the heart of climate crisis. He argues that climate change fiction is not taken seriously: 'the mere mention of the subject is often enough to relegate a novel or a short story to the genre of science fiction'. Ghosh insists he is not devaluing sci-fi by this statement; rather 'it is as though in the literary imagination climate change were somehow akin to extraterrestrials or interplanetary travel' (7). He suggests that the avoidance of climate change in 'serious' literature is because the topic upsets the rationalism of modern life into organised narratives. The realist novel and readership, born of bourgeois notions of moral betterment, conjures up realities that are actually a 'concealment of the real'.

While there are more novels on climate than Ghosh acknowledges, and because of this his ideas are disputed, nevertheless his argument resonates in the context of a neoliberal museum 'concealing the real' by maintaining neutrality on climate change to support the status quo.

World endings

Other fictional narratives contain a fatalism that can be interpreted as a psychic need to use guilt to maintain a sense that humans are in control. Cultural theorist and philosopher Slavoj Žižek (2017) identifies this psychic need in relation to the adoption of the Anthropocene concept: 'There is something deceptively reassuring in this readiness to assume the guilt for the threats to our environment: we like to be guilty since, if we are guilty, then it depends on us, we pull the strings of the catastrophe' (2017). His analysis identifies a profound species narcissism. Guilt control might reveal a psychic level to the museum's broad take up of the Anthropocene thesis. A phenomenon in museums that might be explained as a form of exhibition guilt. And this might extend to a knowledge and power relationship attached to admitting climate change through its presentation in museum displays. Is institutional guilt then, as the psychology Žižek observes, a counter intuitive mechanism of control?

A behavioural quandary arose from my perhaps wayward reading of Barbara Kingsolver's climate novel *Flight Behaviour* (2012). Lepidopterists undertake field research in Appalachia into the monarch butterfly's relocation and impending extinction. The scientists gather, document and interpret data on the dead and dying creatures. The accuracy of their speculation that the butterfly is disappearing validates their research, amounting to a strange passivity toward extinction rather than a move toward activism.

This kind of nuanced response to the tragedy of climate change and species extinction touches on the complexity of grief and guilt relations toward the more-than-human.

It seems that the global north engages scientific knowledge as an ordering of the world even when the order is chaotic. There is a distancing from things under the microscope rather than perception of intimate entanglements. Yet we are microbiomes and holobionts – diverse collaborative organisms made up of trillions of bacteria, viruses and fungi. What is required is to be able to think the reality of our holobioticism. A recent study acknowledges this awareness underscoring the potential role of the microbiome in perception and action, to such an extent that 'The way we think about *how* we think may need to be revisited' (Jake Robinson and Cameron 2020, 5). Such thought requires storytelling that engages with our multi species becoming with the matter of the earth.

In new stories of *homo sapiens*, the minerality of life is wiser to encounter than construing soils, minerals and fossil materials as stuff to exploit. It is a form of geocide to degrade the materials that make up the human *Umwelt*. While this can seem difficult territory to push into western logic, we can be inspired by ideas and provocations that challenge common-sense categories and make it hopeful to think differently. As Deleuze (1994) writes: 'Is it when we do not recognise, when we have difficulty in recognising, that we truly think?' (138).

Holobiotic entities, existents and agencies in art, fiction and fabulation all chip away at the fixed assumptions we call common sense. These stories often take western logic into the world-making familiar to First Nations Peoples – ways of walking with the earth that cannot be thought when thinking is locked into the reasoning of western humanism.

The world-making in N.K. Jemisin's *Broken Earth* trilogy (2015–2018) does just this. Her narrative presents the way the world ends, for the last time. Fifth Seasons are eras where earth becomes untenable for life, except for stone folk who exist forever, sort of. There are Acid, Choking, Boiling and Breathless Seasons; the latter is described thus:

> Location: Nomidlats, Sathd quartent. An entirely human-caused Season triggered when miners at the edge of the north eastern Nomidlats coalfields set off underground fires. A relatively mild Season featuring occasional sunlight and no ashfall or acidification except in the region; few comms declared Seasonal Law. Approximately fourteen million people in the city of Heldine died in the initial natural-gas eruption and rapidly spreading fire sinkhole before Imperial Orogenes successfully quelled and sealed the edges of the fires to prevent further spread.
>
> (Jemisin 2015, 452)

In the invented reality of *Broken Earth*, Orogones are a special being who can control energy, particularly through moving the ground with their sessapinae.

They are oppressed by the race of humans, who are known as the Still. The world-making takes us into the type of category-crossing that the rational modern self finds it difficult to enable because it entangles thinking with monstrous and abject others, and as is well evidenced: 'Enlightenment Europe... tried to banish monsters. Monsters were identified with the irrational and the archaic' (Tsing et al. 2017, M5).

Having surveyed the literature on the various ways of framing climate change, Elizabeth Boulton (2016) finds a 'complex, multilayered, and busy space' with 'a growing chorus of argument that the arts and humanities must be better mobilised for this task given their expertise in "making meaning"' (773). But it is difficult territory to traverse. Boulton surveys a range of framings, such as fear messages and catastrophe narratives, as largely ineffective – 'framing information in terms of loss and uncertainty can be detrimental to motivating action' (774). She notices that 'attention is turning to fundamental philosophical questions' within the climate change literature citing Palsson et al. who propose that 'traditions of Western thought have reached their limits, and that it is the question of "how to think" in a climate-impacted era that needs primary attention' (Boulton 775).

Reviewing climate communication, Boulton picks up on Morton's view that current thinking is redundant to the task of responding to climate. 'Humans' preexisting, hardwired, or deep frames are so mismatched to the new era, humans must rely on other ways of knowing, such as sensing and feeling – or attuning' (Boulton 2016, 780). One of her conclusions however is that Morton's own writing can be obscure and this may make his ideas impenetrable. But dealing with the present crisis and going into an increasingly uncertain future it seems inevitable that ideas are complex as they do not correlate with common sense. Solutions are multi nuanced, and here again we find the active role for museums – to be lucid and visceral communicators of difficult and tacit knowledge.

Adoption of the Anthropocene

The scholarship for museums to disentangle includes debates and controversy on the efficacy of the Anthropocene concept, encountered in the previous chapter. Dutch chemist Paul J. Crutzen and biologist Eugene Stoermer (2000) advanced the thesis that humans are a geological force on the planet. As they note the idea is not new as in 1873 Italian priest and geologist Antonio Stoppani wrote of an anthropozoic era, rating mankind's activities as a new telluric force.

Nicholas Mirzoeff (2018) considers the Anthropocene is inherently a racist conceit; for what kind of man is this Anthropos? He suggests the era is more accurately a White Supremacy Scene and that science representing geological time as stratigraphy is aligned with the way that racial distinctions are stratified. The alignment between climate change and the carbon economies of fossil capital is a racialised relation affecting those least able to resist catastrophic physical events.

Anthropocene's entry into scholarship and discourse has taken aback stratigraphists who are wary that global awareness about environmental change is being meshed with stratigraphy, which they explain is a practical tool that solves geological problems (Autin and Holbrook 2012, 61). Anthropocene rhetoric across science communities and environmental movements is forging ahead when it is not clear there is sufficient evidence 'to define a distinctive and lasting imprint of our existence in the geologic record' (60). It is a daunting task to formalise a stratigraphic ranking of an epoch or age according to stratigraphic codes: 'a stratotype that records a continuous, preferably marine sedimentation record and separates the Anthropocene from underlying units needs to be identified and correlated into the global time stratigraphy' (61). Autin and Holbrook question whether a stratigraphic measure is a suitable way to gauge such change.

I wonder that sedimentary measures of leachates, plastiglomerates and nuclear fallout do not qualify as a geological gauge of change? Regardless, the concern of stratigraphers does highlight the difficulty in making connections between different areas of expertise in such a way that aligns with traditional values of that field. But this move is what psychoanalyst and ecologist Félix Guattari championed as an early seeker of how to produce better ethical subjectivities. He discerned this would occur by 'shifting from scientific paradigms toward ethico-aesthetic paradigms' (1995, 10).

In the tool box of approaches available to museums to attune to an ethical aesthetic paradigm in climate governance is Donna Haraway's adoption of the Chthulucene as a way to acknowledge the wondrous ordinary relationships between things. Haraway adopts the Chthulucene to bridge hopeful entanglements of species and technologies, and to voice, like Mirzoeff, the arrogance of naming the modern era after man: 'Living with and dying-with each other potently in the Chthulucene can be a fierce reply to the dictates of both Anthropos and Capital' (Haraway 2016, 2).

The term Anthropocene has been described by museum scholars as a keyword in the sense articulated in the 1970s by Marxist cultural theorist Raymond Williams (Isager et al., 2021, 88). Williams observed that at particular junctures keywords become everyday experience and integral to forming industrial capitalism. That the Anthropocene thesis is popular at this late capitalist juncture is a phenomenon that environmental and literary critic Rob Nixon (2018) aligns to a shift in attention in the 2010s from the shock of 9/11 toward fear of planetary extinction. Understanding *homo sapiens* as a 'fused biological-geological force' framed a new situation to that of the twentieth century where human occupation of the planet was deemed geologically trivial (Nixon 2018, 2). As Ellsworth and Kruse (2013) also acknowledge, 'something is happening to the ways that people are now taking up 'the geologic''' (6) an event surveyed in their edited collection on making the geologic now. They recognise the need to take up an expanded understanding of geology:

> Of course, "geology" continues to reference rocks, tectonics, and bare forces of our planet, including deep time. But it is taking up new associations as people struggle to understand and meet new and unprecedented material realities of Earth and life on Earth.
>
> (12)

In areas of the humanities concerned with intersections (rather than connectivity mantras), there is frustration at the limited criticality of much Anthropocene discourse. Elizabeth DeLoughrey (2019) observes that a 'lack of engagement with postcolonial and Indigenous perspectives has shaped Anthropocene discourse to claim the novelty of crisis rather than being attentive to the historical continuity of dispossession and disaster caused by empire' (2). What she is alert to is that the apocalypse *has already happened*, and this requires that Anthropocene discourse in relation to climate change takes notice of theories, approaches and stories that have been already told.

> There is an unprecedented production of climate change books written by geologists, in which an undifferentiated 'man' has a starring role in the history of the planet, causing speculation about the behavior of the species in the past and dire warnings about its actions in the future. These environmental morality tales are, of course, allegories of a universal masculine subject who is not subject to cultural, historical or sexual difference. When Anthropocene journalists insist that the term 'man' is gender neutral, it seems as if the decades of work about context and difference in the humanities never existed.
>
> (DeLoughrey 2019, 11)

Will Steffen (2013) observes that the first attempt to define the Anthropocene appeared in the newsletter of the global change research program International Geosphere-Biosphere Programme following its introduction by Crutzen and Stoermer. It caught on. He notes that in pulling together various projects the Anthropocene 'became a powerful concept for framing the ultimate significance of global change' (2013, 487). The general and non-specific meaning of the term is iterated in debates about when the Anthropocene begins as a 'golden spike' that can be measured in the rocks; does it commence with agriculture, with industrialisation or with the atomic age? With any of these 'spikes', attention is drawn to the destructive impact of humans, which returns the question to which humans? Naming an epoch or age accrues a great deal of power to whomever those humans are, and in doing so blames a nebulous 'humanity' for the damage of fossil fuel capitalism while also assuming 'man' can geo-engineer and fix anything. It is the type of hyper-humanism that continues to pursue the notion of an ideal human, not a transformed and more modest critter.

X-factor

In their research of museum exhibits on the Anthropocene, Isager et al. (2021) consider the widespread uptake of the term and that it is 'among the rare scientific concepts with sufficient X-factor to cross over from its scholarly field of origin into virtually every other academic discipline, into the worlds of politics and the arts, into mass media and into the worlds of museums' (88). That a range of climate change exhibits are themed on the Anthropocene tells us something about the museum. The term signifies scientific pedagogy, but in a vague kind of way that is non-specific and imprecise. As Nixon (2018) observes,

> From Germany to Australia, Switzerland to the United States, curators are staging ambitious Anthropocene shows that range, in mood, from the celebratory to the despairing, from the earnest to the antic. The Age of Human is making itself felt in modest galleries and mega art shows, from the Venice Biennale to Art Basel Miami.
>
> (7)

As Nixon argues, with 9/11 no longer a priority, museums had to address the issue of anthropogenic impact on the earth. They could no longer approach climate disruption as an incidental topic, so they adapted the Anthropocene by taking a neutral position on climate change.

Naming a 'human age' is surely as contentious as pronouncing a universal museum, and for a similar reason. 'The universal "we" ... is the most powerful narrative of the Anthropocene' writes Marco Armiero (2018, 131). What about the *anthropo-not-seen*? asks Marisol de la Cadena (2019), as she draws attention to world-making processes that do not separate the human and non-human (40). Art historian T.J. Demos (2017) too views the Anthropocene thesis as a universalising discourse of text and imagery that obscures the accountability behind the mounting eco-catastrophe. The use of visual imagery has been integral to the process of conceptualising the Anthropocene, yet the images are far from transparent or direct and there is no effort to educate audiences. He argues that the scientific popularising of Anthropocene imagery enables 'the military-state-corporate apparatus to disavow responsibility for the differentiated impacts of climate change, effectively obscuring the accountability behind the mounting eco-catastrophe and inadvertently making us all complicit in its destructive project' (19).

Subtle forms of avoidance in dealing with the vital issue of responsibility for the climate are apparent across museums, for example, in the presentation of Anthropocene Studies at Pittsburgh's Carnegie Museum of Natural

History. The museum created The Anthropocene Living Room, for visitors to 'relax and learn about this epoch'. The website frames the experience as a comfortable pedagogical engagement while noting: 'Impacts that will be present in the geological record millions of years from now, and today represent a crisis of sustainability and what many are calling a planetary emergency' (Carnegie). Does the museum agree that it is a planetary emergency? And who is responsible for the calamity? And what must be done?

T.J. Demos (2017) does not want to give the attention to humans that the term Anthropocene suggests, while Nigel Clark suggests that naming the Anthropocene can be useful. He cautions that 'What is vital for critical thinkers in the humanities and social sciences to recognize, ... is that the scientific thematization of the Anthropocene is as much about the *decentring* of humankind as it is about our rising geological significance' (2014, 25). It is the latter interpretation that museums can adapt as they move away from anthropocentric ordering of the world toward acting on climate by entangling the inorganic and organic and in doing so creating a new framing of inclusion.

While in many ways inadequate, Anthropocene is now commonly used to describe human impact on the earth's physical systems and does admit a need for cross-disciplinary discourse around anthropogenic global warming. That the cross over is seen as a dumbing down of individual disciplinary practices presents a role for museums, that is, to bring into clear view insights that generate a deeper way of understanding and valuing the difference of non-human things as a difference with the self.

Death of the world

The didactic staging that is common to geological encounters in museums communicates the knowledge that stratifies the earth. But what is the visceral lure of an affecting encounter with soils and rocks and their processes of change and transformation? Like other museum visitors, I am drawn to geological collections and displays in museums; the lure is a feeling that these objects are wildly out of place and time. There is a durational effect of existence outside human temporality. If I gaze at a rock for my lifetime, its material change will be imperceptible. I move momentarily into planetary duration. Rocks mess with my relationship with who I am. Artist Ilana Halperin (2015) captures the experience of duration through her relationship with Iceland's Eldfell volcano:

> Standing on the volcano, I thought about how Eldfell and I are now almost forty. I wondered about returning to Eldfell when we both turn fifty, sixty ... How, while we both share our lifetimes now, that will only continue for a certain amount of time, and then Eldfell will go from a human life scale – thirty years old, forty years old – to a geological time scale – 150 years old – 1,000 years old, 800 million years old.
>
> (80)

With burning forests and rising seas, moving the frame of reference away from controlling the Earth's physical systems is imperative. Shifting the tenor of human-lithic entanglements can reset the western imaginary. That people are actants in the lives of rocks is sensible thinking for First Nations Peoples and yet profoundly difficult for western subjects for whom rocks are exemplary of nonlife. Stone dead is a powerful metaphor.

So where are new regimes of perception that attune to non-human others? How do museums shift from being bound to dualistic thinking? Perspectives that may appear dystopian can suggest paths to more inclusive relationships with the earth. An example is Andrew Culp's contextualising of the philosophical project of Gilles Deleuze – known as the philosopher of difference – through the proposition that the world needs to be destroyed (2016). Some will balk at this, for aren't we urgently trying to save the planet? However, Culp's 'death of the world' confronts what thought must become to resist the destructive world of unfettered capitalism.

Culp updates Deleuze, recognising that becoming structurally resistant to the excesses of capital requires rejecting the joyful connectivity mantras found in superficial readings of Deleuze by 'rootless rhi-zombies, dizzying metaphysicians, skittish geonaturalists, enchanted transcendentalists, [and] passionate affectivists' (1). What irritates Culp is that the role given to difference and otherness is fuzzy among those who cite its importance. Here I find an interesting alignment with museums given their collection, research and exhibition of different forms of 'others'. Culp's 'death of the world' is a thought experiment to move outside the conceptual framings of anthropocentric humanism, like that found in museums. He qualifies:

> This is not a call to physically destroy the world. The Death of God did not call for the assault of priests or the burning of churches, and the Death of Man did not propose genocide or the extinction of our species. Each death denounces a concept as insufficient, critiques those who still believe in it and demands its removal as an object of thought …
> (Culp 2016, 66)

Other theorists of materialism, vitalism and the posthuman adopt similar stances. Literary philosopher Claire Colebrook (2014) in *Death of the PostHuman*, a book on extinction (ironically titled Vol. 1), frames self-extinction as 'the capacity for us to destroy what makes us human' (9). In lieu of the catastrophic ecological, political and cultural crises shaping futures in the 2020s, destroying what makes us human before this version of humanism destroys life on the planet seems increasingly responsible. But where are lucid and understandable perspectives to transform thinking toward such a proposition to be presented? It is here that museums that are self-reflexive about their legacy play an imperative role. Notions such as death of the museum or the dark museum offer a bridge to new knowledge not dystopian or utopian endpoints.

As with Culp and Colebrook's extinction concepts, an ending can be hopeful. The goal is moving beyond the hubris of anthropocentric assumptions of human exceptionalism. The end of the human as a subject position separate and independent of material and physical systems provides a path outside the logic of fossil capital and its requirement that *homo sapiens* must survive at all cost. In fossil capital, it is necessary for humans to persist for where else are markets for commodities?

As it stands, institutions supporting economic growth hold more weight than the fragility of life on the planet: 'We increasingly suspend the *thought* of our fragility for the sake of ongoing efficiency' (Colebrook 2014, 13). And so, Colebrook asks, 'how might we imagine a world without organic [human] perception? Is there such a thing as perception without a *world*? (23). Colebrook is referencing Martin Heidegger's infamous claim that a stone has no world. She poses:

> Can we imagine a mode of reading the world, and its anthropogenic scars, that frees itself from folding the earth's surface around human survival? How might we read or perceive other timelines, other points of view and other rhythms?
>
> (23)

The perceptions forwarded in the approaches of geo materiality, geontology and geo poetry create imaginaries where soils, pebbles and mountains carry perspectives that do not align with the logic of geology. The regime of perception to move toward, in support of proper living on the planet, is that we terrestrials exist inside the geologic, inside the weather, moving amongst shifting, sifting sediments that are not within our control. Undoing former perceptions is relevant for museums that stage displays on the Anthropocene. Displays and interpretations that assume the centrality of Anthropos need to decentre this notion to enable images of difference, such as what it means that humans are irrelevant through the perception of non-human inorganic actors and existents.

Hylomorphism

The ethical 'turn' for a geo-inclusive museum is to not only prioritise past and present exploitation of the earth, but to do so through an understanding of how this exploitation emerges from the abilities of *homo sapiens*, most profoundly how the western imagination thinks and understands the world. Turning again to philosophy, an approach is to understand Aristotle's notion of hylomorphism, a term derived from the Greek words for matter (*hule*) and form (*morphe*).

Aristotle believed that objects are made of four elements – earth, air, fire and water – and that these change into one another to make forms. Every physical object is a unity of matter and form, and can take on any form. In this idea, Aristotle distanced himself from Plato's theory of forms, which purported that there are universal ideal forms that exist apart from the material world.

In hylomorphism, making something begins with 'a formless lump of "raw material" and ends when form and matter are united in the complete artifact' (Ingold 2012, 432). Tim Ingold writes that 'form came to be seen as actively imposed, whereas matter – thus rendered passive and inert – became that which was imposed upon' (432). Like Ingold, critics of the hylomorphic model reject the form-receiving passivity of matter, including Deleuze and Guattari who reference the anti-hylomorphic stance of philosopher Gilbert Simondon. Simondon showed that dualism and hylomorphism kept understanding of the process of individualism in a zone of darkness. He opened a new way of thinking about the individual subject as always incomplete and more than a unity or identity. Things, including human being, are always in a state of becoming the other.

The fluid effect of a never-ending process of becoming human led Deleuze and Guattari to adopt the concept of metallurgy to reflect the inadequacy of thinking of materials as passive and inert. Metallurgy is a process of following materials, a process that does not accord with Aristotle's fixity of materials. It is what artisans and artists do; they go with the flow of materials in a fluidity addressing that a material cannot be pinned down to a concept. What, for example, is the concept of gold? Of oil? In recent theories of new materialism all things are a matter of movement, fluidity and mobility between elements. Here we find the ontological basis requiring humans tread more carefully on the earth.

In philosophising on relations between materials and thinking, Deleuze and Guattari distinguish an alternative nomad science or nomadology. This works on the assumption of the fluid formation of matter in various contexts that create ways of becoming and smooth spaces that do not accord with the dualistic thinking that supports the structures of late capitalism. Western institutions rely on a fixed set of concepts that have become common sense, but this fixity is not how things are. Smooth space is not Euclidian space: 'matter-flow can no longer be cut into parallel layers, and movement no longer allows itself to be hemmed into biunivocal relations between points' (Deleuze and Guattari [1980] 2004, 410).

These ideas can appear densely abstracted as they escape common acceptance of the status quo. With the idea of fluidity, 'the meaning of Earth completely changes' as instead of extracting constants from matter there is engagement with a 'continuous variation of variables'. This does not mean a fall into the irrational, which can happen with any science, but rather reflects a science made from '*following* the flow of matter' (emphasis in original Deleuze and Guattari [1980] 2004, 412).

Rosi Braidotti (2011) develops nomadic thought as her method to move beyond the destructive fixity of patriarchal dualism and human exceptionalism.

The space of nomadic thinking is framed by perceptions, concepts, and imaginings that cannot be reduced to human, rational consciousness. In a vitalist materialist way, nomadic thought invests all that lives, even inorganic matter, with the power of consciousness in the sense of self-affection. Not only does consciousness not coincide with mere rationality, but it is not even the prerogative of humans.

(2)

Nomadic thought is a type of unity that rejects form and matter hylomorphism to enable merging with the earth in ways that respect the agency of matter; here is Haraway's odd kin composition, and Barad's proto consciousness of matter. Braidotti (2011) develops Deleuze's concept of becoming machine as a meta(l)morphosis to create space for a subjectivity or way of thinking and feeling that is an ethical immersion in the current technological environment. A new mode of relation with the materiality of metals, elements, and ores.

My optimism is that museums adopt becoming-earth as good practice. A transformation from the 'matter as form' interpretation of objects toward nomad thinking that communicates the fluid agency of inorganic and inhuman existence. Museums as proof of the possibilities of geo-inclusion. Currently however there is not much fluid thinking around the amassed collections of the 55,000 museums in the world. The project is to repurpose collections in ways that de-centre the human. Museums can harness their capacity to confront the damage that has been caused by institutions, like themselves, remaining neutral across the decades on the causes of anthropogenic climate change. It matters that plastics are removed from sale in museum shops and that artists and activists are let in. But it is more than this; it requires philosophical renewal. And this confronts the dilemma that the museum has become a shiny commodity. It needs rehab, and in this challenge perhaps can buddy that other institution compromised by neoliberalism, the university.

Adaptation

In suggesting geo inclusion as a way to critique the take up of the Anthropocene thesis, other ideas, such as 'globalisation' and 'the world', become problem concepts to explain and evaluate. Like 'Anthropocene' they are keywords, 'where the problems of its meanings [seem] inextricably bound up with the problems it was being used to discuss' (Williams 1976, 15). Ted Toadvine (2018) says that 'nothing is less obvious or less certain than that we know what we mean by "world", that we know who or what "has" world, or even that there *is* a world at all' (57). We need to think 'the world' from the view of being present with the local actants the term hides from our view – soils, leachates, coal. To lay bare our grounding.

Globalisation is similarly difficult. For how global is globalisation? (Saldanha 2017, 235). But like 'the world', while globalisation is hard to pin down, don't we need broad generalising terms? Climate change is surely a global problem. Isn't the globe warming? Yet the term global is confused. Bruno Latour (2017) suggests that globalisation refers to 'two opposing phenomena that have been systematically confused' (12). It creates an embrace of multiple viewpoints yet also means a single, 'global' vision. Hence, 'It is hardly surprising that we don't know whether to embrace globalization or, on the contrary, struggle against it' (13). If the concept embraces that there are multiple viewpoints then this suggests something local not global. For Latour, the supposed single vision of globalisation is actually the modernising front of an obscure elite. In this context, globalisation is a tactic of the Capitaloscene and is not only not equal to the task of climate emergency; it supports an elite that no longer even pretends to care for the inequities of catastrophic climate change.

The ideal of a global vision sits with the way that capital lulls people into a green mentality that occupies their attention, so they don't attend to the more complex reality. The very idea of the earth as a 'globe' prepares its social formation for colonisation (Saldanha 2017, 235). Toadvine (2018) describes the impact of globalisation's gesture to a general equivalence as catastrophic, 'insofar as it inspires a proliferation of means and ends that are ultimately oriented toward no final end, no ultimate goal other than their own continued expansion and proliferation' (600). This is not what is needed to deal with an emergency; rather the Terrestrial needs to sense real 'events and moments that compose our quotidian experience, and … deepen our respect for the present'. I discern here a healthy Deleuze-Barad ontology; the Terrestrial as being or life that is not independent from the strata that compose it, the view that 'life only emerges contingently and locally from geophysical processes' (Saldanha 2017, 234).

Arun Saldanha apprehends that the popular imperative to embrace global interconnectedness is unhelpful and certainly not 'oneness'. 'The technoutopian articulations latent in the term "globalization" itself have become enmeshed with New Age philosophies and environmentalisms and are driven by a yearning for wholeness and finality' (Saldanha 2017, 236). In an often-scathing polemic of such movements, Leigh Phillips (2015) rejects this mode of 'green anti-modernity' as a manifestation of neoliberalism rather than a threat to this world order. In order to resolve the climate crisis, Phillips (2015) writes:

> It is time to abandon the pessimism, crippled ambition and human-hating of Heidegger, Adorno, the Counter-Enlightenment, the postmodernism academy and the primitivist left to return to the essence of socialist humanism: a celebration of our species' proven capacity for moral and material improvement.
>
> (254)

I do not disagree with the inadequacy of New Age and other environmental fundamentalisms, but Phillips' embrace of socialist humanism should be a socialist *anti*-humanism. It needs to embrace the inclusion of materials and physical systems as actants for without thinking in terms of radical alterity any brand of socialism is Sisyphustic. Removing the sovereignty of human agency is a fearful prospect but the way to think a new inclusion or ecology through admitting the thresholds of alterity. As an adaptation to climate change, this is the function of the museum.

Scientists warn that strong climate change adaptations are urgently needed; there is no sufficient binding international agreement to cut emissions. The COP conferences have not achieved the required aim. Tim Flannery (2020) notes: 'The climate emergency is now so dire that future COP meetings must focus on far more than reducing emissions' (152). Such events require adaptation. The IPCC posits a number of reasons why adaptation is essential given that climate changes are unavoidable (IPCC, Climate Change 2014 Impacts, Adaptation, and Vulnerability). It is the unpredictability of the changing climate that is an immediate threat. It is apparent that money spent now will be more effective and less costly than emergency fixes but short-term thinking disables this common sense. It is known what is needed but there is not the psychological will to support coherent, holistic, political approaches that are regionally focused on adapting institutions, infrastructure and human populations. A paradigm shift is needed by museum leadership to communicate the limits of human anthropocentrism. The museum needs to think as a Terrestrial. As walking rock. With odd kin. This is the new inclusion as climate adaptation.

5 White geology and displays of material power

The discovery and exchange of rocks and minerals led to collecting activities becoming a focused endeavour, and by 1755 geology was acknowledged as a subject in its own right. Through geology, the world became increasingly known and knowable with an absence of already extant First Nations knowledge and material thinking. The power or force arising from instrumental use of the material world, Kathryn Yusoff (2018) refers to as the White Geology (4). Through reflection on collections and past practices, museums can counter this absence, to which they contributed, directing visitors to observe what is absent in the schemas of White Geology. Museums are institutions with a tradition of dealing with contradiction, and can address the twinned racist discourses of geology and western humanism as these dislocated the knowledge and care of Country of First Nations Peoples. Through a restored ontological awareness, we might better understand the vast scales of the lithic, and become nuanced about our own cohabitation with earth's actants.

More than an objective field of knowledge, the geologic realism Kathryn Yusoff (2019) explores is 'a praxis of materiality and thought in the geneology of Western metaphysics and racial capitalism and a site of possession in the Anthropocene' (1). Inherent to the power and knowledge relation of White Geology is defining what and whose existence is deemed important. Yusoff's project is to acknowledge an alternative inhuman geography that is able to identify and counter the exceptionalism implicit in naming a geological epoch the 'age of man'. In this context, the museum finds itself in a peculiar position; it both epitomises the knowledge and power relation that is instrumental to White Geology *and*, through this relation, is in a unique position to repurpose its legacy.

A way of understanding the knowledge and power relation that underscores White Geology is to trace the development of this relation as a procession of theories based on observation of mining and metallurgy, and through evidence gleaned from rocks and fossils. There are many precursors to the field of modern geology, but we can commence with two early influencers of the scientific method, Georgius Agricola and his 1556 book *The Nature of Metals* and Francis Bacon (1561–1626).

DOI: 10.4324/9780367741945-6

Metallurgy and alchemy

A divergence in geological knowledge separated the art of alchemy from the logic of 'science' through the influence of Georgius Agricola's (Georg Bauer) 1556 book on mining (he latinised his pen name). *De Re Metallica* (*On the Nature of Metals*) was widely used as a catalogue and guide to refining and smelting metals for over 400 years. Meticulous annotated woodcuts illustrate the text including early furnaces in which iron ore was smelted, workers engaged in activities in and around a furnace – at the heath, crushing ore with a mallet, and roasting ore to remove impurities. The first English translation from the Latin in 1912 was by Lou Henry Hoover, a Latin teacher and wife of Herbert Hoover a mining engineer (and later American President).

Agricola's book is a remarkable document of mining and metallurgy in the sixteenth century. It focuses on mines and processes of mining in Germany but provides a summary of earlier work including those of Pliny the Elder. Iron was used in Greece from about 1200 BCE, reached France and Germany in 700 BCE and in Britain came to name an 'iron age'. Agricola adopts observation and logic, the rudiments of the scientific method, to consider the pros and cons of processes, including alchemy. As a practical guide based on personal observation of metallurgy, his work was understood to supersede alchemy although he is necessarily diplomatic in his discussions as his patrons were cash-strapped German aristocrats and monarchs who would try anything. Hence Agricola (1556/1950) does not dismiss the ancient process, while observing, 'we do not read of any masters [of alchemy] who became rich' (xxviii).

There is recent attention to the material thinking of alchemy with an awareness that it is a more complex pursuit than has been popularised: 'Alchemists have been obscured by misleading classification and a haze of rationalist condescension' (Gaskill 2021, 16). In popular culture, alchemy is associated with dark magic, and this has given it a skewed reputation. It has always been an intellectual approach rather than a discipline, with alchemists seeing themselves as philosophers, and having 'a penchant for reading, writing, making and doing all at the same time' (Nummedal 2011, 330).

Alchemy was a serious intellectual inquiry bringing together materials in an early modern form of correspondence that informed the likes of Isaac Newton and Robert Boyle. Malcolm Gaskill (2021) argues that while alchemy might seem 'a blind alley … it remained attached to and energized by a miscellany of ideas, in constant motion, continually reinventing itself' (16). He finds in this a continuing legacy reflected, as an example, in the development of COVID-19 vaccines that 'proceeded not just from solid expertise, inspired by lab work and rigorous testing, but also from an older kind of curiosity and courage' (1).

One of the things that alchemy does, and that does not follow with modern geology, is not to separate intellectual thought from material flow, from the processes of the artisan – 'the distillers, miners, goldsmiths, and apothecaries whose tacit knowledge and experience of waters, metals, and minerals, as well as the equipment and processes necessary to work with them, constituted alchemy's practical foundation' (Nummedal 2011, 331). It functioned as a flexible and unique inquiry into the material world. As alchemy was not bound to a guild it was accessible to all social classes linked, through its engagement with materials, to the 'artisanal and commercial culture of early modern European cities' (336).

While there is interest in the interdisciplinary engagements of alchemy, Nummedal (2011) notes that a disconnect remains between literary scholars and historians of science. A key interest for the museum and the inclusion of inhuman geology is that alchemy 'often had multiple valences simultaneously' and in doing so tells us something about the siloed nature of contemporary knowledge. A separation of knowledge into fields and disciplinary areas that are fixed so that boundaries between unlike things cannot be breached.

Science and scripture

That the world could be properly known through reasoning applied to evidence was advanced by Francis Bacon. In a startling novel *The New Atlantis* (1624), Bacon envisages Utopia based on an institute devoted to empirical research and investigation of nature. In heralding the scientific method, Bacon was an inspiration for the founding of the Royal Society in 1660. He repudiated Aristotle's notion that experiences from the senses present things as they are to our understanding, arguing that images in our mind do not render an objective picture of true objects. It is necessary to free thought from false idols that are productions of the human imagination and nothing more than untested generalities (Stanford 2013).

It is not that Bacon rejected the biblical Creation. Rather he held that society has to ensure that losses called by the Fall are compensated for by man's increased knowledge. His argument is that it is not wise to confuse the Book of Nature with the Book of God, since 'the latter deals with God's will (inscrutable for man) and the former with God's work, the scientific explanation or appreciation of which is a form of Christian Divine Service' (Stanford 2013). By 1700, rock-collecting Anglican clerics at Cambridge University held this in good stead, believing that empirical evidence of the earth would 'replace blind faith with proof of Christianity' (Cook 2003, 92).

English clergyman and rock collector Thomas Burnet sought to reconcile the gradual formation of the earth with scripture. His *Sacred History of the Earth* (1690) provided a linear history of earth from Genesis and Noah's flood to a succession of biblical events culminating in the Last Judgement, when earth becomes part of another universe. Burnet believed that biblical events could be proven by evidence drawn from collections of material objects. His geological theory, from doing just that, was that time was linear rather than the fixed events of Creation, and this thought gave tacit permission to commence an empirical inquiry into the true date of Genesis. (It was held at the time that the Creation happened on 23 October 4004 BCE, a date derived at by Irish Archbishop James Ussher, around 1650).

That time is not fixed was the reasoning behind a new logic of the earth's origin. The mineral collection of Burnet's student John Woodward led to debates on the origins of fossils, previously believed to have been created at the same time as all other plants and animals. Woodward proposed, however, that following the Flood, fossils reformed from plant and animal remains. These enlightenment thinkers were not secularists – a distinguishing difference with humanism – they were universalists, with man at the top of the ladder of being, beneath God. It was an optimistic age, and all things were there to be discovered; it was Woodward's objective not to collect exotic or eclectic objects but 'to get a compleat and satisfactory information of the whole Mineral Kingdom' (Cook 2003, 95).

By 1755 geology was listed separately in Samuel Johnson's dictionary of English words. In 1759 the British Museum was established as a repository of natural history collections, including the rocks, minerals and fossils of Sir Hans Sloan and other collectors like Sir William Hamilton. Hamilton, an English diplomat based in Naples, published an influential book on volcanoes based on his observations of the eruptions of Mount Vesuvius and Etna. He discerned volcanoes building up the land, and lava forming basalt. Hamilton's character is *The Volcano Lover* (1992) in Susan Sontag's novel, which profiles Hamilton's passion for collecting, including the rocks and Greek vases he sold to the British Museum. His desire to accumulate knowledge through the possession of evidence includes the volcano, when no object could surely be less ownable. Sontag's work is an observation of White Geology, reflecting Hamilton's collecting and material power as inextricable from his class, race and patriarchal privilege.

The classification of geological artefacts was evidential truth that kept giving until by the end of the century geological theories were advanced that no longer conformed with scriptural geology. By 1800 the Linnaean system of classifying the mineral kingdom was refuted by an order proposed by Abraham Gottlob Werner, a German geologist at the famous Freiberg Mining Academy in Saxony. Linnaeus considered the mineral kingdom was the third kingdom in nature. Interested in mining he believed that ores grew in mountains and hills were formed when a rock was transformed into ore by a chemical process. This was part of Creation: 'All this places anyone who reflects on it in a position of awe at the Omniscient Creator's disposition of our globe. Thus speak the stones, when all other things are silent' (Linné online http://linnaeus.uu.se/online/history/mineralog.html). Werner advocated however that rocks originated in a primordial all-encompassing ocean and as this receded, it left behind different formations of rocks. His Neptunist theory opposed the new Plutonist (vulcanist) idea that rocks formed through heat at the centre of the earth and were then forced up (Cook 2003, 99).

Werner did not engage with Noah's flood, and nor did James Hutton, a Scottish landowner and physician who became known as the 'father of modern geology' (James Hutton Memorial 1947). Hutton theorised that tectonic uplift involved the constant denuding and formation of new land, and that the earth was older than 6000 years, and not going to end any time soon (Cook 2003, 101). The recurring cycles of physical activity led him to conclude that there is 'no vestige of a beginning – no prospect of an end' (James Hutton n.d.). Hutton's theory was exemplified in the 'unconformity' of the layering of silurian strata and Devonian rock at Siccar Point in Scotland. At a memorial to Hutton in Edinburgh, three boulders with pronounced granite veins signify his observation that rock strata and sediments demonstrate geological processes. The unconformity of different rock strata reached into creation myth: 'Humans found their senses of self and their traditional mythologies displaced in relation to, or not conforming to, the notion of deep time' (Matt Baker and Gordon 2013, 163).

Touch stone

Jeffrey Cohen notes James Hutton's image reproduced on a Geologic Time placard at the National Museum of Scotland's display of 'early history'. Titled *Beginnings* the display highlights a large boulder that is identified as two and a half billion years old. Here again is the lure of a rock as a material marker of deep time and corresponding brevity of life. Cohen (2015) observed that everyone who passed the boulder was inclined to touch it (189), perhaps in reflection of the agency rocks hold as signifiers that time extends either side of *homo sapiens*. The exhibit that followed this more-than-human touch stone commences the linear history of Scotland with a display of *Early People*. The boulder is positioned as pre-human; it belongs in pre-history, before things get going, while in reality the boulder will exist in one form or another long after *homo sapiens*. The displays are a hubristic approach to the earth by organising the 'natural' world according to a linear history of humans.

Responding to the NMS displays, Jeffrey Cohen (2015) wonders about the use of stones by curators to tell human history: 'they anthropomorphise inhuman matter, so that lithic stories become the tales of a nation' (191). He considers how different it would be if the rocks were rocks; if the rocky material that the National Museum itself is constructed of were the protagonist. Even in 'stories in which we are not the protagonists, the story is still in part about us' (192). Cohen pursues this idea: 'The temporal alterity of stone does not make the lithic any less a collaborator'. There is a 'worldedness' in disanthropocentric stories, in scales that move us away from the centre; 'a sentience that extends into the inhuman, into the life of granite and geodes, a life of embeddedness, artistry and ethical relation' (192).

James Hutton's geological theory interprets an earth that is constantly restoring itself. Building on this notion of uniformitarianism, the Scottish geologist Charles Lyell, a close friend of Charles Darwin, published the influential *Principals of Geology* from 1830 to 1833. A little earlier, in 1815, English engineer and geologist William Smith produced a geological map of the strata of England and Wales that set the standard and outlined techniques used today. It used different colours to show stacks of vertical geological layers. The layering is the method commonly found in museums to represent the land as geological and geographic contours. In 1817, Smith coined the term 'stratification' from observations of coal seams to provide evidence of correlations between regions and across continents (Cook 2003, 102). From this evidence came the terminology of strata that have the names used today: silurian, carboniferous, and cretaceous. Through the cumulative thinking of geology, the world became known and knowable.

But it was already known. There is an absence and diminishment in the development of geology and related sciences of already extant First Nations histories, patterns and experiences of physical systems. Through self-reflexive focus on collections and practices, museums can highlight and counter this absence, connecting visitors to a more inclusive geology that observes what is absent in the schemas of White Geology, which was a force that 'had come of age by the turn of the eighteenth century; the Geological Society was set up in 1806, independent of the Royal Society, and geology was taught at Oxford and Cambridge' (Cook 2003, 105).

A role of the museum in developing the field of geology was to impart the new understanding of the earth by making it visible to visitors. Time could be seen in geological displays in what Tony Bennett (2004) calls the evolutionary museum, 'by accumulating objects from different places, evolutionary museums were able to make new relations of time perceptible' (22). Bennett is interested in how the 'telling of time' in museums,

> in the form of a unilinear developmental sequence provided the conditions for their amalgamation in a totalising narrative, in which the history of the earth supplied the master time which calibrated the histories of life of earth. And those of human civilisations, cultures and technologies.
>
> (24)

In short, the lithic exists in the evolutionary museum to reveal when *homo sapiens* arrived and how certain it is that humans have progressed and will continue to do so. But time is more than one thing; it is differently known and experienced across cultures, and that which is uniform and linear is only so while physical systems are stable. That physical systems like the climate can be contained by being understood is itself an idea, a kind of faith. It is relevant to consider what is happening to western knowledge as physical systems are increasingly neither stable nor containable.

Concerningly, the focus seems to shift to space exploration as a means of extending White Geology by moving it off planet. There has been an expanded vision of the solar system since 1800 with the invention of telescopes increasing the depth of viewing. The geological and the astrological have transformed what could be thought about time and space. Brooke Belisle considers what this transformation means through her reading of a 2010 diptych by Trevor

Paglin titled *Artifacts* (2013). The first part of Paglin's work is a photo of ruins of the first-century Anasazi civilisation set in the cliffs of Arizona. The striated cliffs in the style of an Ansel Adams image convey a white geological understanding of the canyons towering over the ruins with a sublime grandeur. The second part of the work is a vastly contrasting image of space. Faint streaks and dashes in black space are a belt of space debris; 'this band will encircle Earth after humans no longer exist, forming not just a junk pile but a displaced fossil record' (Belisle 2013, 147).

Through Paglin's imagery, Belisle makes a similar observation to media philosophers Sutherland and Hassan, and others, that the move into technological organisation of the sky as sets of coordinates requires caution. 'As geological and astronomical points of view are increasingly displaced beyond human perspective, and aligned with the apparent objectivity of technological imagining, this threatens the imaginative recalibrations new perspectives requires' (Belisle 2013, 148). Technology becomes the message in a digital trans-humanism where the material power of White Geology is expressed as space junk.

Ironbridge

Iron is everywhere, the fourth most common element in earth's crust after oxygen, silicon and aluminium, and is why most soils are red and brown. The stories of iron that circulate in western culture carry the dualisms and metaphysics of White Geology and are how iron technologies are framed in science and industrial museums and as social history at heritage sites. Exhibits and sites focus on the Industrial Revolution highlighting engineering achievements and inventions. An English landmark is the Iron Bridge across the River Severn, which was opened to international acclaim in 1781. Then as now, 'Iron Bridge is the perfect symbol of the Enlightenment, when technological innovation promised limitless possibilities for the improvement of life' (Hayman and Horton 2009, 9).

The Ironbridge region or Severn Gorge in Shropshire was one of the first places to experience the cumulative changes known as the Industrial Revolution. It was here at the beginning of the eighteenth century that coal was substituted for charcoal in the smelting of iron, a significant event reflected in the region's place names such as Coalbrookdale, Ironbridge and Coalport. A famous image of the region by Philip James de Loutherbourg *Coalbrookdale by Night* (1801) has come to epitomise the sublime industrial landscape, showing smoke from blast furnaces clouding the darkened sky.

Following a long period of decline, and because the character of the district was preserved, in 1986 the Ironbridge Gorge was designated by UNESCO as a World Heritage Site (Hayman and Horton 2009, 12). Part of the site – which has ten interlinked museums, including the Coalbrookdale Museum of Iron and museums based in old china and tile factories – is the Blists Hill Victorian Town, a living history museum that opened in 1973. A now familiar critique of such 'immersive' museums is that these sites construct a nostalgic view of the past; they are antiquarian in their historical recreation. Accordingly, Blists Hill has been described as 'promoting a "spectacle" of idealized Victorian labor' and 'as having the authenticity of a film set' (cited in Rutherford-Morrison 2015, 78). Laura Rutherford-Morrison qualifies these perspectives, arguing that the form of time travel they enable through interactivity, simulation and make believe, enables a form of 'playing Victorian' that promises 'visitors the opportunity to experience history in ways that lie outside of traditional museums and academic study' (77).

My view on the debate around the authorised heritage discourse (Smith 2006, 29) is to shift the engagement altogether. Any force to imaginatively construct change that such a living museum may engender can be redirected to entangling visitors with the materials that have enabled not only the industrial sites and social history they are experiencing, but also the trajectory of progress that has led to the climate emergency. As Rutherford-Morrison (2015) suggests, Blists Hill is already an enactment of change.

> history is not presented as a refuge from the tumultuous present or from change itself – if the museum does point to a constant among the transformations that ruled British life in the Victorian period, it is that change is the constant. Blists Hill does not attempt to capture a moment of lost greatness, but the moment of loss itself.
>
> (11)

These sites can be usefully repurposed giving them greater contemporary relevance through epitomising the role that critical heritage can play in communicating a history of industrial development as a history of climate change.

Museums as climate tools

The iron Earth features in science museums staged through the knowledge and power relations of the White Geology. But always, in carving earth into knowable taxa, contours and tables that make common sense of the world in western imaginaries, the climate hyperobject pushes the other way. Iron and metals are known as objects of solid mute knowledge, and changing this perception can be a catalyst to otherwise perceiving the Earth.

By revising the thingness of the objects in collections, museums can bring things together in ways that reveal how human organisation has real impact. Museums can convey that the ontological divide between a mute earth on one side and dynamic reason on the other is an invention achieved through tidy dualisms that are always artificial. It is not 'enlightened' to perceive that the separation of human consciousness from all other material entities is what makes us 'us'. The separation of the ability to reason from a muted earth is an astonishing immodesty exposed by thinkers, artists and scientists who engage with inhuman duration, sentience and perception. Given the museum's role in normalising anthropocentricism, the obligation is to reinterpret realms of the inorganic and inhuman with the same commitment previously given to humanism.

Attention to stories of stones reveal that modern, empirical ways of understanding the lithic are never a clean break, but inherit previous ways of knowing: 'Geology describes the deep past with a vocabulary and narrative structure derived from Greek myth, Genesis, and medieval romance' (Cohen 2015, 82). Jeffrey Cohen discerns that in these early understandings, rocks are a paradox that alert people to the hubris of human exceptionalism. 'Stone is a symbol for that which is bluntly real' yet also, 'a rebuke to our habitual anthropocentrism' (17). Heeding how medieval writers thought of materiality highlights that today the agency of rocks is masked by a modern utilitarian assumption of their passive nonlife. Classical and medieval entanglements give agency to rocks beyond their instrumental use as resources in consumer capitalism. It is a complicated hubris: 'The indifference of the earth to its dwellers paradoxically reinforces human exceptionalism, so that the material world comes to exist for our instruction and use. Yet stone refuses to remain fully set apart, to respect taxonomic distinctiveness' (23).

Classical and medieval lapidaries and collections held great power to heal and to transform human events. Rocks and stones were useful to think with *because* their materiality posed contradictions. Stones are dead, yet human life is entangled in stone; there is a shared materiality and being with stone. Stones appear everywhere in stories as changemakers. In Greek myth, Deucalion and Pyrrha survive the flood to 'repopulate the post deluge landscape by hurling stones (the bones of the earth) over their shoulders' (Cohen 2015, 51). As Ovid (2004) tells,

> The stones started to lose their essential hardness, slowly
> to soften, and then to assume a new shape.
> In a moment in time, by the will of the gods, the stones
> that were thrown
> from the hands of a man were transformed to take on the
> appearance of men,
> and women were fashioned anew from those that were
> thrown by a woman.
> And so our race is a hard one; we work by the sweat of
> our brow,
> and bear the unmistakable marks of our stony origin.
>
> (Ovid Book 1, 25)

There are countless examples of lithic agency within the mediaeval oeuvre, where supposedly 'passive objects of the human gaze' actually reveal that 'rocks have the capacity to organise the humans who look at them' (Cohen 2015, 51).

Through a reclaimed inhuman geontology, we might better understand the vast scales of the lithic, and be more nuanced about our own cohabitation with rocks and soils. Ancient and medieval scholars acknowledge that the existence and duration of rocks present more mystery than certainty. Isidore of Seville (d. 636) believed that words contain reality and derives the Latin word *lapis* (stone) from the fact that when kicked it hurts (*laedere*) the foot (*pes*) (Cohen 2015, 136). As Cohen shows in his reading of Isidore's mineralogical vignettes, the force of what things do holds meaning: 'Isidore foregrounds inorganic liveliness and material interconnection. Stone triggers environmental entanglement' (137). We might learn from Isidore's geological inaccuracy:

> His open and dynamic system tends to proliferate relation rather than reduce things to singular correspondences. He thereby captures something essential about the trajectories of even the most recalcitrant objects. Rocks, bones, curiosities, and texts tend not to stay securely emplaced on a museum shelf or within an encyclopaedic entry. There is always more than that for which a precise description or rigid classificatory scheme can account, some radiant potential to surprise, hurt, obstruct, exceed.
> (Cohen 2015, 138)

Encountering rocks can involve deep ontological questions, or what it is for rock to exist. How is it that a 'rock' has the identification that it does? How we think about rock doesn't change how a rock's chemistry behaves, but does affect that rock's future becoming. In this sense, a rock becomes a relation – a human relation. As soon as rock is designated as such, it is acculturated beyond its essential make-up of subatomic particles. So questions about rocks are historically specific. While rock is what it essentially is, the 'is' changes with historical determinants. The museum can navigate ways of knowing that escape the hubris of White Geology and its transitions, which includes the dominance of digital solutions to human-caused climate crises.

6 Coal and fossil capital

For climate scholar-activist Andreas Malm, the link between coal and machines in the mills of England is key to understanding today's climate predicament, a thesis explored in *Fossil Capital: The Rise of Steam-Power and the Roots of Global Warming* (2016). The impact of fossil capital takes us into the territory of the White Geology, with both conceits reflecting modern history as a regime built on material power that continues today.

Two hundred years ago, Britain began extracting coal from the ground in large amounts to transform heat into motion by means of James Watt's steam engine, an innovation that connected coal fire to a wheel enabling continuous rotation. The success of coal/machine coupling, Malm (2016) argues, is remarkable given that history is 'stocked with inventions petrified into objects of exhibitions or fantasies' (19). His argument is that fossil-based economies did not have to happen, that they are 'the unique creation of Britain' (196). Malm reckons that water power could have been adopted instead of coal. While this insight has been called 'backward facing romanticism' (Vernadsky 2021) it is nevertheless a challenging perspective that is supported by the realisation that cars and plastics also did not have to happen.

Journalist Jeff Sparrow observes that commodity consumption and its various products, notably cars and petroleum, are the result of carefully orchestrated and deceptive public relations tactics used to resist public opposition to damaging new products. The public understood these products would cause environmental pollution and social dislocation.

> … the car culture we take for granted, in the US and around the world, was formed only after a huge struggle by the auto industry, both against other less-destructive transport options and against the environmental consciousness of the public.
>
> (Sparrow 2021, 7)

Malm (2016) contends that now is the time to comprehend and confront that modern history is a fossil economy. While the Industrial Revolution does not share features of the collective transition we must produce today, what is similar is the impediment of resistant vested interests: 'The previous transition,

DOI: 10.4324/9780367741945-7

then, would be not so much a template for the next as a key to understanding and removing the impediments' (17). Malm structures an argument to show that the power of fossil fuels arises from the use of labour to produce the fuel. 'No piece of coal or drop of oil has yet turned itself into a fuel … fossil fuels necessitate … the power of some to direct the labour of others – as conditions of their very existence' (20).

Museums sit at a crossroad in terms of climate governance; they can remain neutral or adopt a clear stance that communicates the history of economic growth that lies behind anthropogenic climate change. The need for a firm stance is highlighted by Russia's 2022 invasion of Ukraine, which has exposed global mismanagement across decades of stalling the transition away from fossil fuels. For museums, the imperative is to comprehend and communicate the impact of not supporting renewable energy sources. While countries are caught off guard by the politics of fossil capital this does not mean that museums follow the same geopolitics. If transition to renewable solar and wave power, biofuels and battery storage had commenced decades earlier with scientific acknowledgement of the effect of emissions on climate change, global warming and impending tipping points, geopolitics in the twenty-first century could be very different.

The exploitation of coal is a material history that can be developed to communicate the meaning of transition and transformation. Coal is a particularly complex more-than-human actant; its agency performs across multiple human platforms. It is the fuel of the modern world, a geopolitical game changer and our planetary heritage.

Coal tales and true

Coal is crushed plant, formed in the carboniferous 360–280 million years ago. The geologic time takes its name from the coal seams that formed from layers of compressed plant sediment, which millions of years later we burn to produce energy. Emissions of carbon dioxide from this event traps heat and concentrated greenhouse gases in the tropospheric layer of the Earth's thin atmosphere.

Museums in regions that continue to extract coal can engage with coal as a modern event alongside acknowledging the climate emergency. The resilience of coal communities that are supported to undergo transition to sustainable forms of industry and employment augers well for regional capacity to change. The museum can be a hub for this transition. It is no longer relevant to memorialise, for example, that mining tragedies common to most mining towns are the necessary price to pay for progress. That coal is a modernising force is not of more consequence than damage to environments and disruption to Earth's physical systems.

To memorialise the mining of coal and collieries above their destructive consequences reflects an antiquarian use of history in the sense that Nietzsche frames the uses of history (Nietzsche [1874] 1997). A community that identifies itself through antiquarianism 'possesses an extremely restricted field of vision; most of what exists it does not perceive at all, and the little it does see it sees much too close up and isolated' (74). In regions where memorialising the past in the present is key to collective and personal identity, the new and evolving is rejected.

The open-air industrial museum at Beamish in County Durham offers a case study. Operating since 1971, the museum presents the relics of north-east England's industrial and agricultural past through replicating this past as a living history. It is described as standing between 'the edifying tradition of the nineteenth-century museum and the "fun morality" of the twentieth-century theme park' (Cross and Walton 2005, 206).

Taken on its own, a solely antiquarian approach to its industrial past by a museum like Beamish supports the impediment to change that Malm calls fossil capital. However, the reality is that coal-based economies have to end. The end should be where museums begin their engagement with communities and coal. Rather than supporting a nostalgic gloss they can show how antiquarian coal histories have been used, and continued to be used, by Big Mining to naturalise links between coal, progress and human futures.

The coal-as-progress narrative carries the legacy held by coal in the English imaginary. Writing in 1865, economist William Stanley Jevons observed that coal commands the Victorian age. For Jevons ([1865] 2013), coal *is* England and England is civilisation: 'the Coal we happily possess in excellent quality and abundance is the mainspring of modern material civilization' (78). Reflecting on consequences should British coal run out, his worry is that 'other nations', especially the United States, will displace England (83). The role of coal in the national character is assumed: '... coal is almost the sole necessary basis of our material power and ... gives efficiency to our moral and intellectual capabilities'. For an Englishman 'who knows the grand and steadfast course his country has pursued to its present point [the future of coal] must be a matter of almost personal solicitude and affection' (79). Here Jevons monumentalises coal, perceiving the inanimate form embodying the power of an individual and the nation. It is about English greatness, and keeping this greatness going – a mythologising that is based on a few embellished facts. At the end of the nineteenth century Nietzsche ([1874] 1997) observed that such monumental historicising can be just as damaging as antiquarianism (71).

Alongside nineteenth-century hyperbole surrounding coal and national greatness are alternative perceptions. In 1873, Antonio Stoppani, who coined the Anthropozoic age, observed:

> England, where human industry is the most fervent, crumbles and caves in, everywhere eaten through by insatiable coal, rock salt, limestone and metal miners. What will happen, when Europe will all be worked through as England, and the whole world as Europe?
>
> (Turpin and Federighi 2013, 34)

An industrial aesthetic is evident in the use of both an antiquarian and monumental historicising of fossil capital that is observable in children's books. Thomas the Tank Engine is only happy when usefully hauling trucks around the fictional Island of Sodor, metaphors for the United Kingdom and the North Western Railway. Though coal was extracted and used since Roman times, the coal industry in Britain was transformed in the nineteenth and twentieth centuries by railways enabling coal to be transported to inland markets and to ports for export. A heritage value attached to the preservation of steam railways in the United Kingdom that Cross and Walton (2005) observe 'fed off nostalgia for stability, eccentricity, craftsmanship, tradition and even Empire' (211). A nostalgia for engineering is aestheticised in a quaint eccentricity that is given to steam locomotives as exemplified by Thomas the Tank Engine.

From 1945, stories of Thomas and his engine friends were devised by The Reverend Wilbert Awdry to entertain his infant son and these tales remain popular today. A significant Protestant work ethic infuses the adventures of the engines, which are characterised as either busy and helpful or lazy. In one tale, Thomas is dismayed when a truck he hauls passes under a hopper too quickly covering him with coal dust; he goes from being useful and happy to disgraced. Without the grace that comes from productive labour he has no purpose.

Thomas, his engine friends and the railway are working assets in the chain of production that builds the nation. This anthropomorphising provides a utilitarian morality tale for the nation's children. It is only as useful resources that objects like Thomas acquire sentience, and the privilege of becoming workmen. That cultural texts for infants are forged around fossil capital and labour is indicative of the challenge to understand our relationship with commodities versus objects as things-in-themselves.

Coal fuels the industrial world that is the legacy of the modern museum. Cultural texts, international politics, mining multinationals, climate denialism and sponsorship are challenges to museums negotiating the way they stage fossil fuels around the reality of global warming. The move from fossils as resources to extract and burn, toward new stories of coal and oil as other-than-human requires a considered shift in imagination. Coal as a thing-itself is a conflicted consideration for museums in confronting the carbon thingness of coal against the enormity of its utilitarian impact as burnt black coal (anthracite) and brown coal (lignite).

As a resource, coal is extracted from its relation as earth, and instead attached to human power to control nature. Coal and power are entwined in the English language in a materialisation of relations based on labour. For Malm (2016), fossil fuels and steam were fetishised *because* they replaced labour (168). The steam produced from burnt coal was the 'ultimate substitute for labour, because it was everything that labour was not. All its virtues were negations of working-class vices' (168). Steam, unlike human labourers, was compliant. Steam is thus an artefact without agency, 'a perfectly docile and ductile labourer, [with] no ways of its own, no external laws, no residual existence outside that brought forth by its owners; it was absolutely, indeed ontologically subservient to those who possessed it' (168). Like Thomas and his compliant engine friends.

While coal has been business as usual for 200 years, the trajectory was never inevitable. Steam engines only ever belonged to the owners of the means of production, 'a tiny minority even in Britain – all-male, all-white – this class of people comprised an infinitesimal fraction of the population of *Homo sapiens* in the early nineteenth century' (Malm 2016, 206). The move from the beginning to later stages of the fossil economy has never been democratic.

> The succession of fossil-fuelled technologies following steam – electricity, the internal combustion engine, the petroleum complex: cars, tankers, refineries, petrochemicals, aviation… – have all been introduced through investment decisions, sometimes with crucial input from certain governments but rarely through democratic deliberation.
>
> (Malm 2016, 207)

The observation should condition the way museums reflect on coal mining and collieries including mine disasters, environmental destruction and the climate emergency. In Britain, where the big event started, much of the evidence of collieries is gone as they are no longer a dominant presence in the landscape. It is only in a handful of museums and industrial heritage sites that the industry is remembered. Consequently, the power relations driving fossil capital and the appalling toll on workers and the environment is underreported and/or simplified. An example of such simplification is the innovation of the miner's so-called safety lamp in coal mining. From the introduction of the lamp there were more fatalities not less as it meant that coal could be worked in underground areas hitherto considered too dangerous (Hayman 2016, 25). Industrial historian Richard Hayman notes that Britain's worst peacetime accidents have happened in the coal industry.

> Disasters can be measured in numbers, on which basis the worst event occurred in 1913 when 439 men and boys from the Universal Steam Colliery at Senghenydd, in South Wales, were killed by an underground explosion.
>
> (7)

Coal mining heritage is too easily simplified by antiquarian nostalgia or monumental historicising when the focus needs to address that fossil capital, through the exploitation of coal mining communities, led to extensive environmental destruction and unstable climate futures that were predictable and anticipated. This is not to argue that coal mining communities be ignored; the language and lives of miners should be given greater perspective particularly as coal cultures are disappearing: 'The special language of the "goaf" (the space where coal had been cut from), the "rolley-way" (of underground tramway) and the "dib-hole" (the sump for collecting water at the bottom of the shaft) no longer means anything' (Hayman 2016, 6).

A review of recent studies on pit closures in Britain reflects that while some mines have become heritage sites, tourist attractions or parks, many have disappeared and with this the past is erased (Sutcliffe-Braithwaite 2021, 22). The founder of Beamish Museum Frank Atkinson sought to ensure the past was not erased. Atkinson did not intend to create a mass tourist attraction, but

> to rescue a representative collection of objects illustrating a way of life in the region which was rapidly disappearing and to encourage the people of the North-East to appreciate that the history of their forbears were worth remembering and something to be proud of.
> (Cross and Walton 2005, 218)

Opposition to his vision came from the political motive to supersede the region's 'old black image' for one of 'new housing estates and clean light industries' (219).

It is timely to revisit the value accorded to industrial heritage sites, which are often berated for idealising and simplifying the past. It may be an opportunity for a living history museum like Beamish or Ironbridge to present the recent past from the perspective of a post fossil fuel era. They can use this opportunity to respond to criticism that they perform more like theme parks than provide accurate representation of industrial communities, and in doing so gain new relevance.

The potential exists for industrial museums and heritage sites to open up new social projects around coal that interpret the history of miners and collieries to expose the politics of fossil capital from colonialism to the current era of late liberalism. Theories of biopower and biopolitics – on the tactics of the modern state to exert influence and regulation – can be refashioned so that the cultural memory of mining exposes not just the exploitation of labour but of the inanimate. Coal miners and coal are not distinct; they are an object of power controlled by the tactics of fossil capital.

Germany

Munich's Deutsches Museum opened in 1906 on Coal Island (now Museum Island) in the Iser River. Promoting science and engineering, the building was constructed of concrete and reinforced concrete, the latest materials. Robin et al. (2014) observe that the original galleries were

> … assembled as progressive histories of scientific and technological development. Rows of objects were arrayed, starting with older, simpler various and, often supported by gifts from industry, ending with the newest and most "advanced" technology.
> (209)

Two centuries later, Deutsches Museum's permanent exhibitions include a display of Materials, Energy and Production that present the topics of mining and energy. The website explains that energy technology is a topic that appears in many other sections of the museum, as the development of mankind is closely linked to energy history. The section on mines includes modern hard-coal mining and ore mining. While it is limiting to gauge the coal experience from the website, there is nevertheless a discernible neutrality in the way the issue is encountered online. 'Climate' is covered in a separate section on Humanity and Environment, where it is the last of five topics: 'A final chapter is dedicated to the global effects of pollution of the Earth's atmosphere (Air) and climate change (Climate)' (Deutsches Museum website). What appears to be absent in this interpretation is firstly the power of fossil capital and secondly coal as a thing in itself.

The exhibition *Welcome to the Anthropocene* (2014) at Deutsches Museum, discussed in a previous chapter, grappled with global warming through a compromised approach to coal mining and its social histories. As in other countries, the transition from coal is multifaceted, which is what enables the Big Mining and energy companies to obfuscate and hinder real transformation. Germany announced that the country is phasing out coal by 2038; however this is unlikely. As the world's largest producer of lignite coal, mining has shaped the landscape and identity of its coal regions. The coal mines in North-Rhine Westphalia are a case in point. Towns, forests and farmland were cleared for the expansion of lignite mines. The village of Manheim on the edge of Hambach Forest was moved, and the old town and most of its forest no longer exist. Ninety percent of the Forest was felled, a tragedy that drew large protests. As is happening elsewhere in the world, the region's energy company is transitioning to renewables. Mining, however, will continue to expand and the coal 'exit' to renewables does not meet the Paris Agreement's 2030 target. The 2022 war in Europe is likely to further curtail such an exit.

Gaining local support for change involves enabling people to envisage the land once coal is gone, with mines turned to lakes and a focus on bringing back biodiversity and wooded areas. 'Increasing biodiversity through more wild, wooded areas and the creation of the lakes are popular ideas, as is commemorating the role mining played in the area through monuments and museums' (Bateman BBC Future Planet 2021). The commemoration role recognises the museum in its traditional role as a maker of a particular memory of the past; it does not suggest critical and activist engagements with the present forces of the fossil economy. It returns to the idea of museums having a legitimising role in re-presenting a particular conception of coal mining as a national heritage rather than extinction scenarios. The uses of history are powerful and can be used to seduce and dull thinking. It does not need to be this way; critical history can expose the harmful and deceptive use of the past and in doing so critique the political role of fossil capital in the present. Museums can be bold and generate critical histories, exposing commemorations of the past that lack engagement with the false storytelling that maintains fossil fuel–led growth economies.

The compromises around fossil fuels have to be exposed not glossed as too difficult. Listening to environmental journalist Jo Chandler outline the failure of media to deal with climate change and the fossil fuel industry, it strikes me yet again as strange that otherwise informed people are unable to fathom the level of urgency she has been reporting for decades (Chandler 2021). Solastalgia is a term for the psychological distress of feeling powerless and lacking control over climate and environmental change. The condition of solastalgia is growing with the accelerated disruptions of extreme weather events and the failure of institutions to act. The 'new' form of solastalgic distress adds to the bio-physiological pathology induced by environmental pollution and toxins. Psychiatrists identify that distressed environments are inhabited by distressed people, and there are now case studies of this in the context of open-cut coal mining, power station fallout and severe drought (Albrecht et al., 2007, 95).

The coal and climate forecast is not new. In 1912, a prescient 'Science Note' appeared in New Zealand's *The Rodney & Otamatea Times* titled Coal Consumption Affecting Climate.

> The furnaces of the world are now burning about 2,000,000,000 tones of coal a year. When this is burned, uniting with oxygen, it adds about 7,000,000,000 tones of carbon dioxide to the atmosphere yearly. This tends to make the air a more effective blanket for the earth and to raise its temperature. The effect may be considerable in a few centuries.
> (Kollar 2021)

In the blog in which this note is cited, Collection Manager Albert Kollar muses on how a visitor to the Carnegie Museum of Natural History might encounter exhibits that enable them to understand fossil fuels. He cites his own modest project of reviewing landscape paintings at the Carnegie Museum of Art galleries to look for documentation of causes for climate change.

> I found many examples based on the use of coal as a fossil fuel for power and coking in steel mills and the natural formation of bio-methane as portrayed in ecosystem landscapes of the industrial age of the middle 19th and early 20th century.
> (Kollar 2021)

It is interesting to consider this initiative reflecting art historian Andrew Patrizio's call for an ecological art history that involves cultural analysis in fields beyond art history. Kollar's endeavour would seem to fit this cross-disciplinary venture given that he is Collection Manager for Invertebrate Paleontology.

There are a multitude of duplicities around coal extraction and processing. Although it is as fossil capital, and not coal as a thing-in-itself, that is responsible for global warming (Malm 2016), coal is nevertheless anthropomorphised as either wondrous or evil. It acquires moral traits, while also perceived as inanimate and without agency. In promotional advertising paid for by the Mineral Council of Australia, coal is conceived as a magical object (Baker 2017a). It is simultaneously marketed as extraordinary and an ordinary chunk of black rock. Andreas Malm (2016) reflects that the entire fossil economy, as the instigator of climate change, has been normalised to such an extent that: 'A person born today in Britain or China enters a preexisting fossil economy, which has long since assumed an existence of its own' (16). The fossil economy's normalised existence is based on coal having no agency; it does what it is told.

The time of coal

The 2021 *Transition Statement on Global Coal to Clean Power* at the COP26 Climate Change Conference states, 'coal power generation is the single biggest cause of global temperature increases' (UN Climate Change 2021). The European Union signed up to the transition, which includes scaling up technologies and policies in this decade and ceasing issuance of new permits for new coal-fired power generation projects (ibid). But COP26 commitments to phase out some fossil fuels are not enough to avoid catastrophic warning beyond 1.5°C. Some countries did not sign the agreement, including Australia, which argued that market forces will drive reductions in emissions. Prime Minister Scott Morrison was disinclined to attend the conference and only made an appearance under duress. As one pundit notes, 'Australia's ambition for COP26 was to get away with it. To do as little as possible' (CNBC 2021).

The politics of coal in Australia is the focus of scholarship and rigorous investigative journalism (Pearse 2009; Hamilton 2007; Beresford 2018; Wilkinson 2020). It is apt discussing the Australian situation as coal as 'the dirty politics of climate change' (Hamilton 2007) reflects global difficulties in energy transition. Coal mining in Australia operates through the power wielded by business interests and politicians who facilitate private control over public mining resources.

> From the mid-1990s, a campaign began that sought to promote the continual expansion of the coal industry as being synonymous with the national interest while, at the same time, downplaying the threat posed by climate change and its flow-on effects to the Great Barrier Reef. The effort to establish this pro-coal regime involved one of the most extensive campaigns of propaganda and lies ever mounted in Australia. Exposing this power network and changing the narrative about coal have been the goals of the climate movement.
>
> (Beresford 2018, 58)

As the planet's largest coal exporter (in 2018) what happens in Australia is significant and reflects the power base that museums can challenge wherever such climate change politics operates. In 2013 the Minerals Council of Australia incorporated the Australian Coal Association initiating what is known as the era of Big Coal. Beresford concludes that Big Coal is actually a shadow government in the state of Queensland (153). The Board of the Minerals Council of Australia includes international companies with major interests in coal: BHP-Billiton, Rio Tinto, Xstrata Coal Australia, Anglo American and Peabody Energy. The global movements of shadow governments are in-house affairs and include a revolving door of people who move between jobs in government and industry.

A business relationship exists between Australian coal miners and India despite Indian 'coal baron' Gautan Adani's poor environmental record. India mines coal for domestic use but is also one of the world's biggest coal importers. Adani imports cheap coal, mostly from Indonesia, to fuel his energy business and is involved in the large-scale deforestation of Indonesia for this coal. Environmental damage, lack of business ethics and cronyism have found Adani under investigation for various poor practices; however his empire, like those of other mining barons and companies works through governments that operate beyond accountability. Adani's relationship with Prime Minister Narendra Modi 'is central to the rapid rise of the Adani Group and its grand plans to develop a major private power-generating company based on imported coal' (Beresford 2018, 35). Shadow governments operate globally, which means that Adani pays virtually no company tax for its operations in Australia, where it owns the Queensland Abbot Point coal port (Beresford 2018, 53).

Museums and protest

Rejecting fossil fuel funding is a meaningful action for museums that have previously accepted funds from fossil fuel companies and been beholden to governments that support these industries. Rather than accepting corporate sponsorship from carbon producers museums should actively disavow such corporate activity (Miller 2015). Protests against coal billionaire David Koch sitting on the board of the American Museum of Natural History in New York was soon followed by his resignation; although having donated 20 million to the museum his name continues to be highlighted by the AMNH. Greenpeace sourced IRS (Internal Revenue Service) information by the Koch Family Foundations, which shows these foundations 'have spent $145, 555, 197 directly financing 90 groups that have attacked climate change science and policy solutions, from 1997-2018' (Greenpeace.Org).

The relationship between fossil capital and museums must be interrogated by museums for what it was and is, a wrong and improper practice. In the same way that Donna Haraway (2004) critiqued the AMNH in the 1990s for not exposing the masculine tropes of war and bravery explicitly extolled in the Museum's founding panels, the turn is now to the way fossil fuels have and continue to be monumentalised and mythologised. All museums can take a hard line on this, avoiding the requirement of sponsorship deals, implicit or otherwise. These deals require that museums remain neutral in climate debates, and that they operate within positive antiquarianism and monumental narratives of history that frame a future of technological innovation.

As there are already climate change movements afoot the topic of climate in a post fossil fuel museum sponsorship environment should not be difficult to deliver. A question does perhaps arise that if there is recognition might this feed rather than counter a communal interpassivity around climate change? Will visitors feel: 'I go to the museum to encounter the causes and impact of global warming and this conspicuous act is enough'. If this response is anticipated maybe what is required is that museums tell stories outside the familiar and assumed ways of knowing the world.

In the context of climate, defamiliarisation strategies might go some way to overcome the passivity associated with museums as comfortable memorialising spaces. As memorialising spaces, it is common for nostalgia to prevail rather than deep thought. Though of course the experience of deep thinking is difficult to assess. On this I find Ernst van Alphen's (2001) insights useful on a genre of holocaust art that appeared in the mid-1990s. He has studied why such art is affective given the special solemnity expected in representations of the Holocaust and Nazi Germany, for example, Polish artist Zbigniew Libera's *Lego Construction Camp Set* (1996) which pictures a toy gas chamber. The idea that such a work makes is that the Holocaust cannot be taught along traditional pedagogical lines, it cannot be mastered, it should be toyed with. Creating a familiar solemn image of the Holocaust enables viewers to comfortably emote the empathy expected of them. In contrast, Libera's images 'invite viewers to envision themselves in a situation comparable to the real situation. Identification replaces historical distance and mastery' (172). The artwork of a Lego gas chamber envisions the museum or gallery visitor as a perpetrator, so 'we are put in the shoes of the victimisers not of the victims' (173). The point van Alphen (2001) makes about the affectivity of defamiliarising an event is that 'soliciting partial and temporary identification with the perpetuators contributes to an awareness of the ease with which one slides into a measure of complicity' (178).

It may be that this mode of critical intervention is most productively set alongside a memorialisation of a lost or destroyed past. In our climate-changed world, it is the desire to attain comfort, the production of which is fuelled by carbon-intensive energy, that has led to the current situation. Hence, we may soon yearn nostalgically to memorialise such a past. But this ought not deter critical voices that engage with this nostalgia or memorialising as part of the problem in the first place. Perhaps it is this predicament that marks the human tragedy of our time.

Making art as fieldwork is increasingly a method of action to raise awareness of climate change and its disruptions, particularly at sites that are relatively accessible and where fossil fuels are extracted. Being in the field makes it difficult to avoid the incongruity of consumer lifestyles provided by the extracted minerals. Erika Osborne teaches at the Art and Design School at West Virginia University. She describes the impact for her students of first-hand experience of mountain top removal mining in West Virginia, which has the reputation of being a state of extraction. Mountain top removal mining blasts the tops of mountains to reveal coal seams and in doing so destroys the mountains. 'In mined areas of Appalachia, the biodiversity of some of the oldest mountains in the world are obliterated instantly, habitats are immediately lost and clean water sources are buried or left highly contaminated' (Osborne 2013, 62).

Osborne's students meet local activists and view the devastation caused by mountains that have been literally moved by the technology of extraction. The effect is profound.

> When asked what they think of the experience, many students find themselves without words as they try to make sense of what they have witnessed and what it means for them. Although silent at first, the experience ultimately leads to discussions about how to process the encounter not only as an artist, but also as a member of a contemporary culture than survives on the energy created by such practices.
>
> (Osborne 2013, 63)

The level of impact described by Osborne is meaningful, and it is relevant to consider how museums might generate such encounters for their visitors. And the role art can play in this encounter.

7 Oil utopias and petro-invisibility

Writing on petrofiction in the early 1990s, writer and literary critic Amitav Ghosh (1992) noted that oil was largely invisible in literature. On rare occasions when oil does appear as a theme or motif it is invariably limited in scope. Given this limitation, an aesthetics of oil might be uncovered, Graeme Macdonald (2017) suggests, by comparing recurring motifs in petrofiction so that across countries and time periods, the 'structures of feeling produced by oil modernity' would emerge (291). What Ghosh and Macdonald reflect from the observation that the real situation of oil's disruptive impact is mostly hidden is a genre of storytelling responsive to the 'consumptive petrotopia of neoliberal modernity' (LeMenager in Macdonald 2017, 291). Rather than fictional storytelling as a critical history of oil, there is instead a recurrence of petrotopic discourse – oil utopia as the progressive force behind the contemporary world.

A similar discourse occurs in museums, overlooking cause and effect relations between oil and gas and First Nations dislocation, environmental destruction and climate emergency. In an evaluation of oil culture in North American museums and fiction, Stephanie LeMenager (2016) finds that narratives affirming petro modernity are particularly prevalent in the oil museums and heritage sites she visits. The paradox of a progressive-destructive logic underpins the global oil complex and requires theorists to pry deeply to apprehend the affective side of petroleum subjects – the role that oil plays on the American imagination. These petroleum subjects can be identified and evaluated wherever there are petrotopic narratives. The cover of LeMenager's book *Living Oil* is the iconic image of heartthrob James Dean's character Jett in the film *Giant* (1956), drenched in black oil and howling in ecstatic delight following an oil strike on his Texan ranch.

While it is apparent that humans alive in the 2020s are intimately plugged into oil, gas and cars, the relationship of communities to the extraction of 'liquid gold' goes unremarked. There is some attention during oil and gas exploration stages as First Nations Peoples and advocates protest against development, but once this is quelled, extraction carries on relatively unhindered with the support of governments and apathy of citizens and mainstream media. The invisibility of petrol lives needs to be critically evaluated

DOI: 10.4324/9780367741945-8

as, 'we will not make an adequate or demonstrative transition to a world after oil without first changing how we *think, imagine, see* and *hear*' energy in our culture (Macdonald 2017, 292). The complex meaning that oil holds over our subjectivity has to be uncovered if we are to imagine transitioning to a post-oil world.

In public museums, petro-invisibility amounts to a neutral stance on the impact of fossil fuels on climate, while in the oil museums that LeMenager (2016) visits in Alberta and Texas that are sponsored by Big Oil, the messaging is what she calls 'hocus pocus'. Communications are designed to trick, conceal and mislead, by turning a site of oil extraction that has been 'strangely impoverished by its own wealth' (158) into an innovative avatar of the future. The role that industrial museums can play in repurposing the story of oil and gas is not insignificant. 'Regardless of what historians think, or what I think, about bombastic museums of industry, they are staples of American popular culture and contributors to public history' (169). Given the authority entrusted to museums of industry, it is no longer ethical to conceal from audiences the environmental, geopolitical and psychological impact of oil on climate change.

Avoiding a clear stance toward the impact of fossil fuels on the climate, exhibits in museums that represent themes of oil and energy focus on innovation and ingenuity in engineering, robotics and chemistry. The underlying message is that exciting futures await based on sustainable progress through science. The communication delivered through these exhibits and displays is a component in the dynamic assemblage of petro modernity. In local museums and heritage sites, immersive experiences and texts are presented that present history according to the whim of oil companies. A historical timeline at the Ocean Star Offshore Drilling Rig and Museum in Galveston, Texas, exemplifies this. The timeline observes the event of Egyptian drilling innovations in 2000 BCE but not the two world wars or the Vietnam War. 'What the timeline does find historical includes: Barbie dolls (1959), color televisions (1951), Model Ts (1908), the DVD (1995), and diverse innovations in the space, personal computer, and medical industries including the inventions of Prozac and Viagra' (LeMenager 2016, 170). This oddly patriarchal progression is not the critical type of historical knowledge necessary for life and action, but a harmful version of events that does not distinguish between monumentalised oil and mythical fiction. In this misinformation, 'whole segments of the past are forgotten and only individual embellished facts rise out of it like islands' – a version of history that 'deceives by analogies: with seductive similarities it inspires the courageous to foolhardiness and the inspired to fanaticism' (Nietzsche [1874] 1997, 71).

The few petrofiction novels produced during the twentieth century do confront the sudden and acute displacement of communities, whether in Saudi Arabia, Europe or America. It is a commonality of local upheaval and displacement that has become a focus of the energy humanities, a field of inquiry whose 'core rationale maintains that (particularly carbon-based) energy has been abstracted and relatively under-determined in cultural and aesthetic terms, despite being produced, generated and circulated by cultural production' (Macdonald 2017, 292). In taking a stance on climate, a governance role for museums is to replace the zealous attention and misinformation given to fossil-fuelled-futures with a critique of fossil capitalism and what it hides.

The invisibility of oil extraction and the countless conflicts and disasters attending the industry has arisen because oil mostly 'happens' in peripheral communities without a voice to oppose the violence of the petro industry; 'the experiences associated with oil are lived out within a space that is no place at all, a world that is intrinsically displaced, heterogenous, and international' (Ghosh 1992, 30). Ghosh wryly observes that the idea of an epic poem about the oil industry seems ludicrous, that embarrassment about the industry makes it an unspeakable topic, like pornography (29). Macdonald (2017) also remarks that in oil frontier fiction, 'the petro-world appears as if by magic and as magic itself: estranged and unprecedented' (299). Big Mining arrives in remote places with new powerful machines, extracts oil and, in reshaping the land, destroys the original place forever.

The hiddenness of oil spreads beyond the displacement of periphery and First Nations community. In a study of the Rocky Mountain National Park (Boxell and Wright 2017), researchers highlight the centrality in generating mass tourism of petrocapital, cars and road building. 'Once inside the park, tourists not only entered a landscape engineered for auto tourism and other hydrocarbon-dependent forms of leisure, but its wildlife were also intimately shaped by roadways' (120). Like Ghosh and Macdonald, the question to expose is why there has been scant attention to the 'linkages between fossil-fuel wealth and leisure activities' (120). Certainly, carbon energy appeared so abundant and American consumption of petrol was so large from 1945 to 1973 (123) that the materials themselves became irrelevant to a Capitaloscene of highways, modern cars and leisure.

Oil, now

Since BP's 2010 Deepwater Horizon disaster, the oil giant's funding of Tate Britain has generated discussion on methods to expose and resist relations between corporations, the state and public museums. Emma Mahoney (2021) surveys the effectiveness of protest against oil funding and observes the effect changing over time. She advances a cross-institutional critique that operates from within the museum, a wave of activism she observes is now in play: an 'anarchic resistant force has been embodied by a coalition of artist-activist collectivists that infiltrate the museum from below in order to open spaces of resistance within it' (410). Mahoney draws on Simon Critchley's 'antipolitical' interstitial critique to show the efficacy of using museums own resources and capacity against itself, 'its physical infrastructure, its vernacular, its identity and its cultural capital' (413).

An example from among Mahoney's survey of acts of resistance in art museums includes a Liberate Tate intervention that took place at Tate Modern during a Kazimir Malevich retrospective. Developing the event of the display of Malevich's famous painting *Black Square* (1913), the group held up a 64-metre piece of black fabric and invited visitors underneath (Mahoney 2021, 413). The intervention drew on the Tate's adoption of live performance, with the group using this medium to deal with the hiddenness of the oil and culture relationship: 'Tate had repeatedly contravened the Freedom of Information Act, by concealing the details of their financial agreement with BP under "black squares" of redacted text' (413).

Mahoney (2021) observes that the critical intervention provided by the members of Liberate Tate was informed by their background as art school graduates, 'conversant in the vernacular of institutional critique' (416). An important observation that points directly to the significance of supporting art and critical theory in the humanities, and, relatedly, a caution that with the corporatisation of the university, tenured academy is increasingly distanced from teaching undergraduates. The vital function of universities to 'teach' critical thinking is decimated in many universities, a project that sits comfortably with the agenda of fossil capital and its funding of mining and technological research in the academy.

An emergent critique of fossil capital in the 2020s is discernible in TV and streamed thrillers that deal with climate and environmental emergency. The series *Thin Ice* (2020 Yellow Bird) hits a punch as a geopolitical critique of petro-modernity. The series is a co-production between Icelandic, French and Swedish companies. Set in present Greenland, a number of shadow nation states engage in a brutal deception to exploit a newly discovered oil deposit beneath the melting ice. *Thin Ice* presents the geopolitics of oil companies taking advantage of the warming climate they have contributed to creating.

The fiction follows the actual disruption of the climate crisis in Greenland, where warming ice has already caused significant dislocation for First Nations communities:

> The sea ice is thinning to a degree that makes sailing easy for incomers, but hunting impossible for the native Greenlanders. The intricate stages of hardening through which sea ice annually cycles – frazil, grease, nilas, grey – are no longer being fulfilled in many places for the seawater is spiking above the key freeze-point of 28.6 F. When the men cannot travel safely over the sea ice, hunting becomes difficult. Seals haul out further offshore. Bears die of starvation rather than bullets. Inlets and fjords are dangerous to cross.
>
> (Macfarlane 2019, 334)

There is an ongoing struggle to prevent mining in Greenland, yet in 2013 the parliament voted to repeal the country's zero tolerance policy on the mining of uranium and other radioactive material. Greenland has self-government but remains a dependency of Demark and needs to secure an economic base to achieve full independence. The fishing and tourism economy would be boosted by mining, and hence the extraction industry, here as elsewhere, has the small population bitterly divided. The country has a tradition of mining; it is part of Greenland's geological imaginary:

> Political and business elites have long framed their discussions of Greenland's future as part of a geological imaginary in which subterranean resources and their extraction, processing and material transformation provide future wealth and the means for possible independence from Denmark.
>
> (Nuttall 2013, 7)

This is tied to exploration for uranium, which commenced in Greenland in 1955. It is no surprise that in 2019 President Trump proposed buying Greenland.

A pro mining campaign is promoting global warning as a positive event for Greenland as it is making the country greener. Greenland is transforming from an extreme physical landscape to an accessible location that is open for business. As part of this change, Greenlanders are being identified and promoted as a pioneering people for their support of mining. A rhetoric that presents a disturbing contrast with First Nations Peoples elsewhere who are fighting to protect ancestral lands from the destructive impact of mining companies.

In *Thin Ice*, the villain is the United States who sabotage an international climate change treaty led by Denmark that is seeking to prevent mining in the Arctic. Through private oil tycoons the United States falsely blames the treaty's collapse on Russia, while paying off vulnerable Greenlandic Inuit to garner their support. The plot parallels the realpolitik underpinning the exploration and mining of fossil fuels.

The politics and vulnerability of Greenland provides an international business opportunity. Australian resource companies, for example, support Greenland Days at the University of Western Australia (Nuttall 2013, 7). An Australian-owned company, Greenland Minerals, in a partnership with Chinese company Shenge Resources, is seeking rights to excavate the mountains at Kuannersuit in Greeenland for both uranium and rare earths. Its Kvanefjeld mine project is described as 'a large-scale rare earth project with the potential to become the most significant western world producer of critical rare earths' (https://ggg.gl/kvanefjeld-project/). In 2021 however a new government came to power that does not condone the mining because of the fear of extracting radioactive uranium (Gronholt-Pederson 2021). The company argues that changing the goalposts is illegal and continues to fight for an Exploitation Licence to progress the development of the project.

There is lack of knowledge of adaptation to climate change in relation to mining metals in the Arctic, despite Arctic warming being more pronounced than any other region (Tolvanen et al., 2019). Critical evaluation is required to cut through the rhetoric generated as companies in the Arctic update their mineral strategies to secure or maintain a social licence to operate (2). The business of securing a social licence to mine in a region is too often a manipulation of the facts, with experience across the globe revealing that mining companies rarely sustain a community of place. Companies, infrastructure and labour come and go. What has to be masked by mining companies in their cultural heritage strategies to gain a social license to mine in Greenland are Inuit relations to the mountains, where minerals are the veins of the earth. Here as elsewhere, mining the earth, cutting into, removing and changing land forms *kill* the land for First Nations Peoples.

When the politics of oil flow into a museum, a neutral stance on the impact of fossil fuels follows. A neutral stance on extraction is a ploy that does not acknowledge the hidden depths of petro modernity. An example is the messaging at the Carnegie Museum of History. 'As of 2016, the world consumed over 97 million barrels of oil daily', writes geologist Hannah Smith in an essay linked to an Anthropocene website at the Carnegie (Smith 2021). Writing about oil at Titusville and Oil City in Western Pennsylvania where she grew up, Smith considers what the combusting of this oil means for the Earth's atmosphere. She lists various ill effects and the destruction caused since the day in 1859 Edwin Drake drilled the first oil well 'in her backyard'. While Smith anticipates that the industry 'will cause an existential threat to humanity' there is no sense of urgency, blame or responsibility. The day-to-day conveniences of oil are *fait accompli* along with the negative impacts of oil drilling on the Indigenous Seneca Nation, after which the Seneca Oil Company is named.

Smith's approach can be evaluated as transhumanist in the assumption that engineering and scientific innovation will find a solution to combat the climate. It is a common way of avoiding the real issue. As artist Marina Zurkow acknowledges, 'We, all of us who live on the grid in the USA, are soaking in petroleum and wouldn't know how to live, feed, shelter, clothe, or express ourselves without oil-based products' (Hannah and Krajewski 2015, 84).

The ominous environments Zurkow builds in her art she refers to as 'near impossible nature and culture intersections'. Her hand-drawn video animation *Wink, Texas* (2012) loops landscape and atmosphere elements around a beautiful toxic sinkhole. The hole dominates the scene, while in the distant background we glimpse the oil fires that are the *raison d'etre* of the small oil town in Winkler County. So it is that art presents a petro-aesthetic that is too often missing in museums. There are two large and expanding sinkholes in Wink, the result of decades of drilling for oil and gas. Earth scientists explain:

> Wink Sink #1, formed on 3 June 1980 near the abandoned Hendricks oil well 10-A. The second sinkhole...developed on 21 May 2002, centred on

the water-supply well…about 1500m south of Wink Sink #1. The Salado Formation, a thick sequence of interbedded halite and anhydrite, is about 260 m thick beneath the Wink sinkholes. The formation has been naturally influenced by the dissolution of the Salado salt units, but the petroleum activity from 1926 to 1964 around the Wink sinkholes has been suspected to be a trigger that accelerated the dissolution of the underlying bed.

(Kim et al. 2016, 2)

It seems that only deceit enables the smooth extraction and production of fossil fuels. The stories told by oil and mining corporations replace the reality of oil with the rhetoric of petrotopia. To a degree this can be garnered from philosopher Reza Negarestani's docufiction *Cyclonopedia* (2008), which surveys oil power in the Middle East. As perhaps implied in the book's subtitle – *complicity with anonymous materials* – oil in Negarestani's book represents 'the undercurrent of all narrations' (19). Oil is fossil-turned-energy and holds a zombie power; it is 'the black corpse of the sun'. The gesture to personify oil and petrol confronts its impact on thought, and the complex intergenerational power of petro modernity. Oil is fluid not solid; it has a liquid agency that enables it to seep into and saturate the thought of nations. This seems to be Negarestani's theme, which is useful to consider when seeking to make visible the agency of exploited geopolitical actants like oil and gas. The fluid materiality of the thought of countries, cultures and religions amounts to a hyper obsession in the book with fossil fuels taking a demonic form (in the Middle East) and a sociopathic petromania (in the West).

In Negarestani's tale, the cult of Akht is the obsession of an Iranian archaeologist and academic. The cult are worshippers of oil for whom 'the flowing source of the black flame' (10) finds substantiation in the Cross of Akht. An artefact with a star shape that embodies 'a buried terrestrial sun which must be exhumed, a rotting sun oozing black flame' (12). A strangely compelling story of obsession unfolds through digressions, to build an assemblage of oil, terror and the geopolitics of the Middle East. The power of oil wealth is both perpetuated and masked by religion and fanaticism. Politics is the puppet master behind the script, which is probably why the book reads like horror fiction. It is not a straightforward read, as befits the hidden depths of the Petrolocene with the philosophy underpinning Negarestani's assemblage including 'nexuses between numeracy, Telurian dynamics, warmachines and petropolitics, models for grasping war-as-a-machine and montheitic apocalypticism, all in connection with the Middle East' (15). This storytelling has a fundamental obscurantism that is an attribute of oil culture the book paradoxically seeks to expose. Despite this, such philosophical probing suggests new ways to approach fossils turned into energy and fanaticism, and for museums raises questions of how they might address and communicate wayward appraisals of the Petrolocene.

La Brea discoveries

LeMenager (2016) offers a more pragmatic encounter with oiltopia through her experience of visiting a number of North American oil museums. She begins in Los Angeles at the George C. Page Museum of La Brea Discoveries. The La Brea discoveries are Pleistocene fossils preserved in the asphaltum (tar) pits above the Salt Lake Oil Field. Before Pit 91 became a scientific excavation site for fossils, from the eighteenth century the Lake Pit was mined. Within the city of freeways, the asphalt and oil became 'safe' from extraction when it became part of the museum, which opened in 1977. LeMenager proposes that this petroleum/large mammal archive offers a contrast to other such sites as 'oil has been decoupled from any clear idea of progress' (145). The site has the potential to become a post-oil museum, which she defines as 'a place where the cultural meanings of petroleum and fuel more generally are on display, as a reflection on the age of conventional oil that we are now exiting' (145). This seems to occur through memorable encounters associated with diorama and habitat groups, in this case a tar pit covered in water on whose bank a large fibreglass Pleistocene mammoth struggles as she sinks, watched by a father and baby.

That the display is an affecting visitor encounter is telling: 'In an era in which screen culture has become so ordinary as to make less of an impression on museum visitors than an apparently authentic experience, the tar pits present a discomfort that signals the "real"' (LeMenager 2016, 146). While climate change is not a focus of this museum, it could be; the petroleum patch is an in-situ ecological feature, 'an ideal site in which to raise such environmental concerns' (148). It suggests that a fossil archive can be an oil field and disembodied energy, or a paleontological site. For LeMenager the site is an indoor-outdoor museum with the sense of an ecomuseum and the spectacle of entertainment. 'In the twilight of the fossil fuel regime and the shadow of climate change, it simply needs to reanimate its vision' (155).

On the contested topic of the pedagogic impact of spectacle, I recall Toby Miller (2015) musing that it may be necessary to incorporate spectacle into museums that engage actively with climate change and the deceptive rhetoric of Big Oil. While often spectacle suggests passive populism, and not real transformation of thought, he nevertheless writes:

> I think the lugubrious hyperrationality associated with environmentalism needs leavening through sophisticated, entertaining, participatory spectacle. A blend of dark irony, sarcasm, and cartoonish stereotypes effectively mocks the pretensions of high art's dalliance with high polluters.
> (2015, 151)

A suggestion of LeMenager (2016) as she considers the potential for the Page Museum is to highlight significant funded research into the pit's living bacteria that are 'progeny of soil microorganisms trapped in the petrol sumps tens of thousands of years ago' (155). Oil bacteria are prepared for the worse that can happen with climate change and will endure past the end of conventional oil. Despite the momentous implications suggested by a new species and branches of life in a post-oil world, the displays of such a micro futurity at the Page Museum were poorly imagined and easily overlooked at the time of LeMenager's visit. On this, she suggests a focus question for the post-oil museum; 'the Earth that humans damaged through our profligate use of fossil fuels just might be saved by microoganisms grown up in petroleum sludge' (157).

Alison Laurence (2022) suggests that the La Brea Tar Pits operate in a boundary zone, where the still seeping asphalt or tar 'blurs the boundary between audience and exhibit and troubles the artificial border between people and other animals' (71). As well as encountering exhibits from the Ice Age in the middle of Los Angeles, one of the largest cities in the United States, Laurence observes that 'People come to walk their dogs, picnic, play lawn games, and gather with friends, tolerating the smell of hydrogen sulphide that emanates from the asphalt seeps (colloquially called tar pits)' (72). She suggests that visitor encounters at the site where extinction events occurred 10,000 years ago, and that continue to occur in the asphalt 'death traps', offer a proximity that affects the urgency of environmental issues in the present. In the sensory experience of La Brea, there is an awareness of geological time that Laurence suggests invites 'visitors to experience a more-than-human past, present, and – because they continue their chronicle even now – an anticipated future' (76).

Tar-nished

The second oil museum and site on LeMenager's tour is the Oil Sands Discovery Centre and Heritage Park in the Canadian town of Fort McMurray, northern Alberta, a visit with insights that commence at the small airport with its promotional images, company jobs and branding. I'm reminded of a similar promotional branding at Perth airport in Western Australia that presents mining corporations as exemplary employers. In Western Australia, where mining is largely based around 'fly-in, fly-out' (FIFO) temporary labour, the social dislocation in mining towns tells a more complicated story of labour. A 2011 Australian government inquiry into the use of FIFO workforce practices received numerous submissions addressing the negative impacts of these workforces ranging from family breakdown and drug use to isolation, lack of housing and decline of community engagement (Parliament of Australia).

At Fort McMurray the primary resource is bitumen, which is a mixture of petroleum, water and sand. At the town's Heritage Park, there are 17 historical buildings where the region's oil, uranium and fur trade industries are presented as a progressive 'heritage' leading into the beginning of the oil sands era. LeMenager (2016) describes her trip north from California to experience the tar sands, the Oil Sands Discovery Centre and Energy tour, sponsored by

Suncor, as 'a hypermediated experience unlike any other' (161). The experience starts with the term 'oil sands', the industry's term for bitumen when 'tar sands' is more accurate. In a theatre at the Oil Sands Discovery Centre, she watched a young attendant stir the bitumen into a slurry, a 10-minute hot water extraction demonstration that runs at intervals across the day. Hanging on the walls of the theatre, picture quilts highlight various aspects of the industrial process, which requires enormous amounts of water to turn the bitumen into oil.

Elevated and sunken tailing ponds at the site hold the toxic water left over after bitumen extraction: 'Positioned alongside the Athabasca River, seeping into groundwater, and large enough to be visible from space, the tailings ponds are the gravest embarrassment of the oil sands industry. Hence the companies direct their strongest rhetorical efforts towards the ponds' (LeMenager 2016, 162). The process is clearly and visibly a significant polluter; nevertheless as LeMenager continues her tour the duplicity intensifies with her guide offering a parable about mining and reclamation efforts: 'We remove everything, clean up the oil spill which Nature left us, and put everything back'. LeMenager writes that 'this equation of intensive mining with cleaning "nature's mess" was offered without irony' (163).

LeMenager describes the playroom activities for children at the Centre as likely to be baffling to kids, as well as a continuing form of duplicity. One activity on a placard labelled Resource asks: 'Oil sand is like filling in a sandwich. True or False?' The answer given is True with the explanation 'The top slice is overburden, oil sand is the gooey filling, and the bottom slice is limestone'. The explanation is both confusing and an obfuscation, with the concept of 'overburden' reducing 'nature' to a hindrance, and a simplification of a complex process that takes the excavation of two tonnes of earth to make a barrel of bitumen (LeMenager 2016, 163). LeMenager concludes that the mediation of the truth of mining bitumen goes beyond the usual corporate greenwashing observing that, 'The corporate fictions of the oil industry have become deeply lived cultures in North America' (167).

The final museums visited are in the East Texas region known as Texas's Golden Triangle named from the 'rich oil fields that developed here after the Spindletop gusher blew in 1901, just south of Beaumont' (LeMenager 2016, 169). The Spindletop-Gladys City Boomtown Museum is an outdoor site with a replica of a derrick. The Texas Energy Museum in Beaumont and the Ocean Star Offshore Drilling Rig and Museum in Galveston are 'museums of petroleum science and industry with strong corporate sponsorship' (169).

The advent of hydraulic fracturing or fracking has provided oil and gas miners with the opportunity to promote their influence in privately funded museums. In these places, fracking for oil is an engineering and geological feat presented through exhibits and 'rides' that are entertaining for children. As the latest extractive technology of fossil capital (Malm 2016), fracking involves drilling into the earth, then using explosions to smash up shale rock that contains gas and oil. The explorations involve a high-pressure mixture of water, sand and chemicals. The relation between fracking and climate disruption is overlooked in museums, with the focus on the geological and scientific breakthrough. This is despite awareness that, as Howarth et al. (2011) explain, the greenhouse gas impact from methane released by natural gas wells makes it dirtier than coal and oil. 'Methane is a powerful greenhouse gas, with a global warming potential that is far greater than that of carbon dioxide, particularly over the time horizon of the first few decades following emission' (Howarth et al., 2011, 679).

Critics of the practice of fracking are concerned about the influence of oil companies on shaping exhibits at the Perot Museum of Nature and Science in Dallas, Texas.

> If oil companies designed the lessons contained in middle school science textbooks, it would be a national scandal. But helping to design scientific displays in natural history museums that host countless school field trips each year: Apparently that's just fine.
>
> (Harkinson 2012)

Josh Harkinson writes about an exhibit at the Perot Museum that comprises a larger-than-life drill bit cutting though a slab of faux rock, and a fracking-themed virtual reality experience known as the Shale Voyager. These exciting exhibits, shown in the Tom Hunt Energy Hall, were supported by a ten million dollar donation from Hunt Petroleum (now owned by Exxon).

The neglect of the effect of fracking on the environment and climate change in the Perot Museum has not gone unnoticed. In 2021 a petition was circulated by the Texas Coalition for the Environment requesting the museum provide a more honest representation of hydraulic fracking. The museum's response to the concern has been to adopt the usual stance of neutrality: the museum's goal 'is not to take sides but rather provide unbiased information that educates, provokes thought, and inspires minds' (https://greensourcedfw.org/articles/group-petitions-perot-museum-add-environmental-consequences-fracking-exhibit).

The stance is similar at other oil and gas museums. The Fort Worth Museum of Science and History promotes online its permanent exhibition of energy and North Texas. The exhibit which opened in 2009 is called Energy Blast, and 'tells the dynamic story of energy resources in North Texas'. The display enables visitors to 'Explore a model drilling site for natural gas; Get the latest info on how Texas is generating power; and Uncover the energy pioneers who put Texas on the map' (https://www.fwmuseum.org/explore/exhibits/).

Bruce Braun envisages fracking in Dakota becoming an everyday and normalised banality with extraction machines that are designed to fit the earth that the machines will ultimately redesign:

> Imagine a horizontal drilling apparatus laid out in a field like a newly discovered fossil, or the machinery required to frack a well abandoned in

the parking lot of a suburban shopping mall. Remove the geological formation from the picture and the technology makes no sense.

(2019, 131)

And plastics …

Citizens are made to feel guilty about their use of plastics, when the culprit is the petrochemical industry who profits from the ubiquity of synthetic polymers we know as plastic. Eight percent of global oil production goes into the manufacture and production of plastics (Davis 2015b, 350). For Heather Davis plastic 'reveals our utter dependency upon petrochemicals' (349). Plastic marks a clear division in the geologic record and does not go away: 'all the plastic that has ever been made, from take-out containers to nylons to IV bags, is rapidly composing a new kind of geologic layer on the earth'; all the plastic produced since 1907, when Bakelite was invented, is still on the planet (Davis 2015a, 68).

Plastic is a new type of rock, with the name plastiglomerate, a fusion of natural materials with plastic debris. Davis conveys the meaning of the artificial construct that is plastic as coming from the fact that it is not of this earth; it does not have an ecology or sense of place. 'There is no local for plastic' (Davis 2015a, 69); 'it has no *Umvelt*, or world that is made in a co-evolutionary fashion' (70). She suggests that this is its effectiveness, it is ubiquitous, without a stable presence. We are not moored to the earth by plastic, as we are by relations to mineral, animals, water and air; rather plastic is all surface. 'It can become anything, infinitely transformable and manipulable to the wills and whims of human invention. Plastic removes itself from a standard ontological category because it, by design and of necessity, is universal' (71). Plastic is similar to the steam engine and to cars, in the sense that these objects did not have to be invented. Plastic too is a profit-driven material; it was developed 'to replace the objects we already had – but at a price and in a quantity that helped to instantiate a middle class defined by consumption' (Davis 2015b, 349).

The neutral stance of museums avoids its relationship to petro-modernity and evades the politics of fossil capital and climate emergency. Museums can explore what it means to refuse the use of plastics in their operations and practices. For museums to eliminate the use of plastics is impossible, but this is the point, that there is no way to extract our lives from plastic; the Internet, for example, relies on 'thousands of underwater and underground cables sealed from the elements with plastic coating' (Davis 2015b, 349). This is one of the features of plastic: it is a sealant, a barrier; it creates the possibility for materials to become 'a monadic identity separated from its environment' (359). In this it sits with an obsession with hygiene, 'it encapsulates the fantasy of ridding ourselves of the dirt of the world, of decay, of malfeasance' (Davis 2015b, 349).

8   Museums inside the earth

Janike Kampevold Larsen (2013) wonders how it is viable to move into a new mode of awareness of the geologic as an act of engaging with present reality that is not bound by the visual construction of landscape. Landscapes always mediate our relationship with the ground. To move into a new mode of the geologic, Larsen conceives the idea of an inverted museum – being inside the earth and looking outward. A gaze from within matter. She is not referring here to the great underground narrative found in stories like Jules Verne's *Journey to the Centre of the Earth*. Rather, in her other way of sighting the earth, from within, she draws on photos and art of rock excavations made to construct roads; these road cuttings, she argues, 'allow us to coexist with a legible trace of deep time as rock faces, stratification, old riverbanks and just plain dirt' (87). Road cuts are not a part of the landscape tradition; they provide a different kind of information.

Perceiving forms from within, as things-in-themselves rather than in the act of being possessed or through the trope of the sublime, is a way to evaluate the tactics of the western landscape tradition of representation. This is the territory of art history and aesthetics, where techniques and methods of composition and framing contain nature in very particular ways. When nature is deemed too immense to behold it is pictured as sublime and in this conscious act brought back under human control. Emblematic is Caspar David Friedrich's painting *Wanderer above the Sea of Fog* (1818) of a lone wanderer standing atop a rocky mountain precipice gazing across the immense landscape. He could not be any higher above the ground. The man's gaze across the land is an act of visual possession.

The European gaze of possession carries into the American and Australian landscape tradition that visually presents the manifest destiny of western civilisation. A recurring trope emerges of the painted landscape signifying the constant improvement of nature. This is a different visual perspective to Aboriginal art that shows Country from within a place or site. The gaze is not from the outside looking in. Country is everywhere and vital, not fixed to one place. David Mowaljarlai describes Country in north Western Australia as a vitality swinging around him. A psychogeography that is 'the result of relentless cultural labour – marking the ground, lodging painted figure in caves,

DOI: 10.4324/9780367741945-9

determining sight lines to other sacred zones, bouncing sounds off rock cliffs' (Gibson 2006, 23.4). This is not a 'mystical' ability, but an embodied experience in which the boundary that demarcates a human body from a rock body does not end at the surface of the skin (Baker 2017b).

Museum collections that made geological knowledge visible influenced nineteenth-century artists who saw themselves as visual scientists. Their landscapes were informed by the new interest in geology, botany and meteorology inspired by journeys of exploration notably Alexander von Humboldt's expeditions to South America. 'Von Humboldt's scientific project, which encompassed or outright invented entire fields from physical geography to meteorology, were anchored upon his training as a mining engineer, which is to say geology' (Fox 2013, 43).

The new geological theories mediated artists' relationship with the land. Art curator Elizabeth Johns (1998) observes that 'the new worlds of north and south America as well as Australia became primary evidence of the theories of uniformitarianism – the convention that earth had been modelled over the millennia by the same forces discernible in the present' (36). What was formerly painted to be a 'natural wonder' becomes an image of scientific evidence 'such as rock formations, lakes, waterfalls and high mountains, all of which revealed the ongoing formation of the earth'. Visitors to museums came 'to marvel at the physical evidence of … the antiquity and physical power of the earth' (36).

They still do.

Moving into a new mode of intimacy with climate, rather than perceiving climate as something immense and outside of everyday awareness, requires that museums critique the power that the gaze of the landscape tradition holds as the common-sense way of seeing the world.

Don McKay seeks this new mode of awareness in geopoetry – a form of geological speculation that shares a space with poets. This is a way to breach the normalcy of ways of seeing, while enabling a sense of astonishment. McKay suggests that 'the practice of geopoetry promotes astonishment as part of the acceptable perceptual frame' … it is 'a bridge over the infamous gulf separating scientific from poetic frames of mind' (2013, 47). What I consider McKay seeking is the poetic as a way to mediate perception of being from within a vital, living earth. 'Geopoetry does not convert the otherness of nature into the sameness of humanity' (51).

McKay suggests the experience available in geopoetry can be found in romantic poetry when wilderness becomes the other, rather than a harmonious encounter with Nature. He cites Wordsworth's experience in *The Prelude* of a looming shore.

> The huge cliff
> Rose up between me and the stars, and still
> With measur'd motion, like a living thing
> Strode after me.

The geopoetic gesture enacts the cliff as a forceful agency that cannot be assimilated, that 'does not live like living men' and where 'the scrim of humanism is torn aside' (McKay 52). McKay does not reduce this encounter with the geological to the sublime, which is a humanist tradition that contains the immensity of nature. Rather, geopoetry enables new encounters with time and space. These encounters do not accord with the linear organisation of prehistory followed by human history, evidenced by geology conceiving rocks as time stamps and stratigraphic timelines. For McKay, geopoetry enables thinking to go to a place where there are no humans or representatives of the homo genus.

> ... and we realize that the ladder extends back through periods and eras to the Ediacaran, and that even at this point we've covered only half a billion of the planet's four and a half billion years. On the one hand we lose our special status as Master Species; on the other, we become members of deep time, along with trilobites and Ediacaran organisms. We gain the gift of de-familiarization, becoming other to ourselves, one expression of the ever-evolving planet. Inhabiting deep time imaginatively, we give up mastery and gain mutuality.
>
> (53)

J.K. Larsen finds a new mode of awareness of the geologic not through geopoetic articulation but by moving inside the Earth. This provides a perception that is intriguing but not reliant on astonishment. Rather, geological duration is cut through and becomes visible as a scar in the present. This is a pragmatic grounding of the landscape tradition based on revealing the ways industry, infrastructure and development are encounters with the land.

Tectonic melodies

Contemporary artists engage in lithic entanglements that the landscape tradition has diminished. Stone sculptor Emily Young writes:

> ... in my studio a conversation comes into play between me and stone, based on me: human wilful, biological, short lived and wanting to engage, and stone: ancient, alive in electro-magnetic ways and communicative once we have established what we have in common: we are fellow entities of the planet, both of us, human and stone embodying its astonishingly rare history.
>
> (Bowman Sculpture n.d., 9)

The assemblage of the artist and her materials departs from the notion that stones are static. The idea that rock is stone dead is an ontological illusion that marks the distancing of the modern subject from the gritty assemblage that is conscious thought. It is as if the ground beneath western feet must not move in order to provide a solid base for thought. Yet this is wrong thinking once we perceive life as energy and transformation.

Artist Robert Kettels captures this notion in a photographic still of his vertical body hovered above the desert (robertkettels.com). In reality he has jumped into the air and the photograph captures him frozen mid-jump. The stillness is both true to what we see and an illusion. I'm interested in how such cultural texts manifest the paradox of fixing stasis upon the ever-moving, churning earth. It is all compost, and re-composition. Responding to paradox will increasingly become an adaptation as we find our bodies emplaced inside and amongst the climate in these warming times.

A recent festival in West Java responding to the environment around Mount Tilu involved a marriage between different types of rocks, a ritual to enact harmonious relations with the 'other'. Music performed on stone-based instruments included a tectonic set of melodies on volcanic rocks by Swiss musician Simon Berz (Aw 2020). Cultures and artists who listen to hear geological forms are attuned to the circular presence of stone and earth as ancestors. Much can be gleaned from knowledge that has been masked by the landscape tradition of western humanism, including grief as an understandable response to the removal of stone ancestors by the activity of mining exploration, extraction and processing.

In the Western Australian Pilbara, countless rocky outcrops carry images in the form of petroglyphs mostly created by pounding into rock using stone tools. It is estimated there are more than a million of these images, suggestive of thousands of years of occupation, which dispenses with the colonial expediency that prior to Europeans the region was unoccupied. It is the incommensurability of First Nations knowledge of the Country with that of mining development that frames the destruction of the area.

Further East, across the state border into the Northern Territory, Elizabeth Povinelli writes of the *thatthere* of a 'place' called Tjipel. She confronts the impossibility of framing in western logic how the Indigenous Belyuen people perceive where Tjipel begins and ends. This is because Tjipel 'does not refer to a thing but is an assertion about a set of … obligated orientations' (Povinelli 2016, 100). Povinelli (2016) seeks to convey the manifestation of nonlife beings or geontology.

> Where does she [Tjipel] begin and end – where the sands accumulate to maintain her breasts or further down shore where they drift off to sea? Are the oysters and fish and mangrove roots and seeds and humans, who come and go as do the winds and tides, *karrabing* and *karrakal*, part of her no matter where they may stretch or travel?…She is not, in other words, in any self-evident way an organism.
>
> (99)

Museum and heritage studies that engage with material flows take the affective and emotional component of a visitor's encounter with objects into consideration. There is an understanding that everything is material culture – emotions, sounds, stones, light, tides, winds. These studies acknowledge that in the western landscape tradition the sense of sight continues to be privileged over other senses as a mode of understanding and that sight and language are inextricably bound together. We employ language to communicate what we see, so that, as Susan Pearce (2010) describes:

> it is on this basis that the modernist world, with its characteristic episteme, has been built. Within this world objects behaved themselves. They did not shift their intellectual shapes or change their places in the received scheme of things, or mean different things to different people, or different things to the same person at different times. But we have always felt that these things were not true; they did not match what happens to us and how we feel about it as we live our lives experiencing our material world.
>
> (xix)

It is realised that emotional value is garnered from the memory generated from felt recognition for an object and that museum objects can be memory in material form (Witcomb 2010). The affect is the visceral feeling that is generated from such an encounter, a feeling that can transform thinking. Museum studies that engage with memory and materiality sometimes take into account the emotional presence or aliveness of things but this is tentative as it moves the scholarship into the realm of feelings. The realm of feelings is not associated with objective and rigorous pedagogy. Material thinking in museums can explore giving greater value to the perspectives of materials themselves as a practice that ranges into encounters with the other as unrecognisable yet something of our making, and undoing.

Let concretes speak!

A brilliant example is a darkly comedic script titled *Concretes Speak* (Harkness et al. 2018), a play in one act that gives a voice to concretes as the central protagonist 'within the epoch known as the Anthropocene' (29). The purpose of the play is pedagogic, with concretes announcing the historical and environmental impact they have had on the construction of modernity. The concretes resent the inattention that is given to their enormous effect and are rebelling.

> … we [concretes] add more to the earth's atmosphere than global air transport does … Despite this, the stories of our production, use, and impact are not as widely recounted as they could be, considering our many entanglements with you.
>
> (32)

In the accusatory address of the concretes the audience is forced to acknowledge how toxins and pollutants are indeed hidden from view. There is also a gesture to the hubris of human endeavour.

> ... we have claimed a voice and are recounting our processes of becoming, our history, our relationship with you, and our role in this, your anthropocentric era. Why should you have the only say here, we thought!
>
> (ibid)

In the tenor of a Greek tragedy, with a Concrete Chorus providing commentary on the action, which involves mixing concrete, the play concludes with a monologue from the defiant Chorus.

> The modern ruins that increasingly surround you – that are disintegrating, sometimes slowly, sometimes dramatically, back into aggregate – these ruins are testament to the fact that we can't live in the eternal present you envisioned for us and so for yourselves. We exist in the human timescale: components gathered, mixes mixed, and slabs poured by individual humans and specific machines. But we also exist in the geological timescale, and particularly within it we are revealed as liquid.
> From rubble we came, and to rubble we will return!
>
> (ibid)

Acknowledging that once-silenced objects acquire an agency arising from that silencing enables the strangeness of new perspectives to percolate. The assumption of human sovereignty over nature becomes the conveyance of the damage caused by this assumption. Museums are actually practiced at this kind of contradiction and can take up a geopoetry of objects in their collections. It is a contradiction that containing artefacts in vitrines or behind ropes also opens that artefact to meanings beyond the quotidian. There is a 'contained freedom' for the visitor: 'to pass through the museum's doors is to cross an ocean to a distant world that can seem very strange indeed in comparison to the one the traveller has left behind' (Dudley 2015, 44). The gaze of the visitor onto the object can provoke unsettlement of that gaze (ibid). But the gaze of the othered object as a relation that cannot be assimilated into an already written narrative is an undervalued kind of rupture. It is a discomfort to common-sense relations with climate that museums can develop and nurture once they move beyond neutrality.

Back to the sublime

What actually happens in museums is often far from geopoetry. In many geological and earth science displays – such as a soil monolith or contoured geological map demarcating stratigraphic layers – the encounter is static with a focus on visualising taxonomic information. At the other extreme, in order to produce 'edutainment' and gamified environments, pedagogic immersion practices in museums utilise digital and interactive technologies. These screen-based practices can enable dramatic staging's of material processes so that an event such as oil extraction or energy production is transformed into a sensory spectacle. I suggest that this use of technology represents a return to the sublime grandeur familiar in the landscape tradition, an oily sublime.

With projection technologies and immersion software widespread and increasingly less expensive, museums can now choose to present earth and cosmic geology as a dynamic collective experience – the big bang, volcanoes and meteorites become encounters in galleries with the fantastic. The dynamism of the universe and the earth's physical systems including climate can be cleverly simulated. Michelle Henning (2006) suggests that what is significant about the popular take up of simulation-based displays is not their illusionism, but the way that they erase what they commemorate (60). Following this insight, simulating the dynamism of the universe erases the dynamic universe. The simulation of the earth's dynamism is ironically a passive engagement. There is also an irony in replicating an earth system when earth *is* the assemblage of materials generating the replication of the system.

The displays at the Wiess Energy Hall at the Houston Museum of Natural Science are themed around energetics, petroleum geology and oil exploration. Much of this is based on simulation. The website says 'The Wiess Energy Hall is the most technologically-advanced exhibition on the science and technology of energy anywhere in the world' (www.hmns.org/exhibits/permanent-exhibitions/energy-hall/).

There is a hyper energetic truncation in the way the displays are described. Here is the wiki link description of what visitors can expect:

> ... a working replica of an offshore drilling rig drill floor, a 15K resolution video depicting the history of energy, the 'Geovator' (a simulated trip into the rock beneath Houston and back in time to the Cretaceous Period), the 'Eagle Ford Shale Experience' (a simulated journey to Karnes Country, TX, to experience the hydraulic fracturing of an oil well from inside the cracked rock), 'Energy City', (a 1/150th scale white model depicting the entire energy value chain brought to life through projection mapping using 32 laser projectors) and Renewable and Future Energy Sources.
> (https://en.wikipedia.org/wiki/Houston_Museum_of_Natural_Science)

Digital projections and immersive experiences are a significant practice at this natural science museum, with activities suggestive of a theme park for oil and gas. The themes of the museum are entertaining experiences of what might otherwise be considered complexly disruptive events ecologically, socially and climatically. The scientific and engineering ingenuity involved in the extraction of oil and gas and its transformation into energy is highlighted. But what of CO_2 and greenhouse emissions, which are invisible or hidden in the themes?

Can visual geopoetry be experienced in displays that do not revolve around digital technology? Prior to COVID restrictions, I visited the Melbourne Museum's geoscience collection in 2020, titled Dynamic Earth. The way the geological specimens are staged creates a pleasing aesthetic experience. Individual rocks carefully positioned in spot lit showcases capture the crystal formations and vivid colours of minerals and gems. Dimly lit spaces within the gallery convey a subtle sense of the underground. The display does not rely on digital spectacle; it offers a different sort of enchantment, perhaps more akin to the lure of a mediaeval lapidary or cabinet of curiosity.

Jeffrey Cohen (2015) writes of premodern displays of 'irradiative gems, agentic rocks, and riotous minerals in jostling congregation' (229). Cohen's study of commentaries on stones in pre-modern lapidaries concerns their powers over humans:

> Some cure diseases, some inflict harm, and others hold powers hidden from human discernment ... Medieval lapidaries are treatises on material vibrancy. Interweaving knowledge derived from Greek, Roman and Islamic authors with meditations upon petric efficacy, they describe what stones accomplish in the world.
>
> (229)

The curatorial intention of Melbourne Museum's Dynamic Earth display is not to promote the magical and healing qualities of stones, although neither is it to present the utilitarian reality of minerals as the crux of modern industry. It is not that the business of digging up the earth is entirely absent, rather that this activity is matter of fact. Alongside the mineral displays but not too close, didactic texts address the economic significance of contemporary mining. A panel reads:

> Australia's Mineral Resources
>
> Australia is the world's largest exporter of raw and processed coal for making steel. It is also one of the world's two largest exporters of iron ore. It is the world's second largest producer of gold and is pre-eminent in exports of titanium and zircon mineral sands.

I reflect on the general tenor of this information. What is noticeable about the displays is what is not divulged about the anthropogenic impact on the land and physical systems of extracting and processing the minerals. The neglect can be framed as indicative of a neutral stance toward the relationship between mineral extraction, fossil capital and climate breakdown. There is a vagueness to the texts that avoids construal of a specific perspective or political orientation.

Later I scroll through the pictures I took to document the visit. There is a distinct dissonance between the aesthetic pleasure in the materiality of the minerals that I capture in my images, and the prosaic economics of global trade the specimens represent. These are different meanings to be had from the museum encounter. But what and where is the museum's story? There are numerous questions that are avoided. To what extent does the aesthetic display highlighting each object's 'finest' qualities cover the real politics, conflict and disruptions attached to fossil fuels and energy production? Does a museum's objectivity represent misinformation? Are there cases where this neglect is deceptive disinformation? Where is the climate emergency and energy crises? Where are the links between nationalism, war and climate? Can a collection be presented without engaging with the assemblages that these interrelated factors construct? Responding to the entanglements these questions confirm are where significant complexities and new relevance are to be found for museums in repurposing their collections.

Meteors

There is a display of meteorite fragments from the geoscience collection at Melbourne Museum. These include pieces of the Murchison meteorite, so called as it struck the planet near the town of Murchison in Victoria. Contemplating this event in the presence of the fragments is a marvel. Older than Earth and the Solar System, the seven-billion-year-old entity, which is perhaps a comet fragment, hit the planet at 10:58 am on 7 September 1969 causing a cloud of smoke, and a tremor (Wikipedia/Murchison Meteorite). The precise time it 'fell to Earth' and the age given it by cosmochemists is a durational jolt. The incommensurability between the brief existence of *homo sapiens* and the geological age of the cosmos realises our profound irrelevance as individuals and as a species. A lesson to draw from meteorite fragments is that stones are very modest.

They are also clever. In a mediaeval book of magic stones, known as the Peterborough Lapidary, a meteor is described as coparius, 'a rock engendered by clouds' with the power to protect its keeper from lechery, lightning and *mysauenture* (misadventure).

> This meteoric rock materializes an intimacy of gales, clouds, rain, bolts, bodies and earth, a coming-into-entanglement. A dense nexus of unpredictable relation-making, stone discloses the enchantment inherent to things, the powers of which cannot be reduced to history, use value, contextual significance, or culture.
>
> (Cohen 2015, 165)

Different cosmologies reflect contrasting entanglements with geological artefacts. The meteorite crater field in Argentina known as Campo Del Cielo (field of sky/heaven) marks an area 18 kilometres long by four kilometres wide. The iron rock formed in space millions of years ago, and 'fell' to earth several thousand years ago, breaking up in the atmosphere. Pre-Columbian people knew the shower of rocks came from the sky, hence the name, 'field of heaven'. An annual National Meteorite Festival continues to be held in Gancedo, the town closest to the site (Viano 2015).

There is a market for these iron objects. Fragments appear for sale online (meteorites-for-sale.com; meteoritemarket.com) with pieces located across the world's museums and collections. A large piece of the iron is recorded in a 1788 Letter to the Royal Society by Spaniard Michael Rubin de Celis. He muses on what the unusual object might be:

> Either this mass was produced in the spot where it lies, or it was conveyed hither by human art, or came hither by some operation of nature. It could not be generated here, according to any known process of nature.
> (Royal Society 1788)

Unable to categorise the anomaly, Celis attempted to blow it up to discover what might be revealed inside, ultimately determining the object to be 'of no value, since it could not be used' (ibid).

More recently, Scottish artist Katie Paterson engages with the same iron body in her work *Campo del Cielo, Field of the Sky* (2012–2014). She cast a fragment of the iron to replicate the object's original form and mass. The cast was then re-turned to space by the European Space Agency (Katie Paterson Artwork). Images document the process – the iron's lunar-like pits and smooth curves, the casting process forming the new piece under tremendous heat and the neo-meteor being transported into space.

The artifice involved in this work is wondrous – a mass of iron that dwelt for millennia on planet Earth that has been cast as a meteorite and taken 'back'. The act of 'back-ness' removes this temporary fragment of the earth from its role as 'iron' and economic resource to be extracted, crushed and manufactured into steel. Here is the agency of rock as observer of the mortality and fragility of life on Earth.

Paterson's cloned iron wants us to think differently through its manifestation as an intersection of human and mineral scales of time. The act of casting and transporting a meteorite off the planet mentally re-territorialises iron. Having been forged by the artist, Paterson's use of iron as a human technology is not dismissed; rather it acknowledges an existence for iron beyond its Terrestrial residence. The art process provides a 'different kind of consciousness ... rooted in contact with igneous chaos' (L.B. Larsen 2014, 1). It is a connection that does not accord with the values driving the Spaniard's curiosity centuries earlier. The thought of 'back-ness' disconnects the found-made object from usual relationships with the lithic.

Paterson's project following *Field of the Sky* is *Future Library* 2014–2114, a public artwork supported by the City of Oslo. The project involves growing a spruce forest and turning 1000 of the trees into books, one commissioned each year beginning in 2014. It is an archive project, as the books will not be read (probably) by people alive today. The project has been associated with climate change as, 'We may not know who future readers will be or what world they will inhabit, but the artwork demands that we take them into account – and it shows how doing so may change what we do in the present' (Bronstein 2019, 121). Michaela Bronstein reflects that not being able to read the books of well-known writers is part of the point. 'We are stayed from consuming something we desire' (122).

The ecological sustainability of *Future Library*, which is housed in the new Deichmanske Bibliotek (Oslo public library) with a special Silent Room to archive the unread manuscripts, has been questioned in the context of Norway's relationship with oil and natural gas. Is the library a luxury purchased by the burning of fossil fuels? Mitch Murray situates Paterson's project as a small part of a much larger urban renewal scheme aimed at redeveloping the Bjørvika shipyard district in Oslo. He writes that the effects of Norway's fossil fuel trade 'will not be offset by a sustainable Oslo' (Murray 2020, 182). Another concern raised by the project's literary futurity is its 'toxic positivity', a critique that argues such attention to the future is part of an American ideology of optimism (Bronstein 2019, 132).

The project does appear to limit the consideration of the forest as forest. While it is deemed about the future, does it engage with climate as an urgency, with the future in the present? There is an irony here that the archived novels, unread for a century, are likely to deal with the burning of fossil fuels, mass extinction, tyranny and the fouling of the planet, given that the first two commissioned authors are David Mitchell and Margaret Atwood. It may be that Paterson's literary archive will be opened and read earlier than expected, as is likely to occur with seed archives, given the fragility of the planet's soils and warming earth.

The Svalbard Global Seed Vault on the Norwegian island of Spitsbergen is a warning of the future in the present in relation to confronting climate change. The concrete seed vault carved into the side of a frozen mountain 130 meters above sea level opened in 2008. The Vault's mission is to provide safe, free and long-term storage of seed duplicates from genebanks and nations, with the purpose of ensuring the world's food supply (www.seedvault.no/). In 2021 the facility held close to 900,000 seed samples. Though situated in what is currently one of the world's coldest places between the North Pole and mainland Norway, the site has nevertheless been affected by warming. The function of the vault is to keep the seeds frozen at -18°C, but with climate change and melting permafrost it is apparent that even the safest site on the planet is not completely secure. There was a melt of the permafrost in 2017, with water encroaching the entrance requiring a major upgrade of the facility between 2016 and 2019.

In contemporary art, an intersection of human and geological time scales can address the reality of existence after the human species is gone. The idea of the posthumous in recent critical theory moves toward this kind of forecast as a way of thinking about relations with the earth. Paterson's post-terrestrial iron event is somehow of this forecast of unknowing while the Future Library is doing something else. The posthumous is interested to theorise what happens when human extinction is conceived, and while this is somewhat of an impossible thought experiment, the posthumous is useful in locating agency beyond human life. It poses questions that 'require thinking beyond the human conditions of existence' (Weinstein and Colebrook 2017, x).

9 Gold on show
The toxic glamour of the yellow rock

Displays of gold artefacts based on a country's treasures, such as the exhibition *Indonesian Gold: Treasures from the National Museum, Jakarta* (1999), are programmed as touring shows. The temporary exhibition and catalogue of *Indonesian Gold* surveyed gold objects in the Jakarta Museum dating from the seventh to the twentieth centuries. Another example of an exhibition based on a treasure is *Gold of Africa: Jewellery and Ornaments from Ghana, Cote d'Ivoire, Mali and Senegal in the collection of the Barbier-Mueller Museum* (1989). The frisson of gold treasure or a gold collection that is treasured by a nation is a status rarely attached to other minerals. The influence that gold holds over human behaviour largely remains a puzzle.

Distinct to the glamour attached to touring exhibitions of gold objects are permanent historical displays related to a region's gold mining past and/or present and future. These museums are tourist highlights in gold mining towns, and are sometimes attached to a disused mine. There are also museums that are based on a collection, notably the pre-Hispanic collection at the Gold Museum (Museo del Oro) in Bogotá, Columbia.

The displays in a museum dedicated to a goldfield provide a history of exploration, mining and development in a region. The origin story (or myth) of the site that led to a rush of miners to a region commences the narrative. Emphasis is placed on the original find of a nugget or seam, as this provides a solid introduction for the story to follow. The finder/s of the 'first' nugget of gold with photographs and a replica of the original artefact are invariably part of the origin story that usually concludes with a positive evaluation of the town or region's relationship with the gold industry.

In the Western Australian Museum of the Goldfields, photographs of early Kalgoorlie show tents at diggings in an arid desert landscape. At first gold was found on the ground and near the surface. Once gold is found at a location, miners go to extraordinary measures to find the metal. The equipment and inventive machinery built by miners inform the narrative of progress in museum displays. Early mining for surface gold used panning and various devices such as a rocker or cradle, a simple machine with a rocking motion used to sift and wash material. These devices used gravity and did not require chemical pre-treatment to recover the ore.

DOI: 10.4324/9780367741945-10

At Sovereign Hill living history museum in Ballarat visitors can pan for gold in the creek. The souvenir booklet says:

> Real gold awaits lucky visitors in the Red Hill Gully Creek! It is salted regularly with fine alluvial gold. Sovereign Hill's Diggings typify the diggings which dotted the plains as thousands of fortune-hunters raced to Ballarat. Alluvial gold, long since eroded from quartz reefs in the hills, lay loose in the sand and gravels of creek beds. Finding this gold was easy. Panning simply involved washing dirt in a goldpan.
>
> (Sovereign Hill Museums 2013, 3)

Alluvial gold often required crushing and cyanidation. The environmental damage of these processes does not find a place in the story of gold in museums. Cyanide is used to treat and extract gold from ore: 'Despite being highly toxic, the cyanide leaching of gold is the most common technique in the exploitation of gold from its (high and low grade) ore bodies' (Ilyas and Lee 2018, 22). Gold minerals that are difficult to extract and 'recalcitrant' are referred to as 'stubborn' and 'headstrong'. These require more use of toxic roasting treatment to recover the gold. Sometimes oxidation treatments are performed prior to cyanidation. Finer gold is recovered using a process of amalgamation; an amalgam is an alloy of mercury with gold. Mercury is a collector metal that brings free gold particles in contact with it to literally dissolve gold (Ilyas and Lee 2018, 21).

In their study of gold metallurgy and environmental impact, Ilyas and Lee describe how processes of recovering gold have been in use for thousands of years. Due to the requirement of large amounts of toxic chemicals and the generation of massive waste, gold is 'dirty'. The processing of gold is not only toxic; it is also energy excessive: A mine of high-grade ore that might contain a few hundredths of an ounce per ton not only requires high energy but may also lead to toxic drainage. And the processing of acid-generating sulphide ores generates sulphuric acid. 'In the extraction process, the uses of hazards like mercury in gold amalgamation and cyanide as the most suitable lixiviant are the biggest threats to the environment' (Ilyas and Lee 2018, 23).

The myth of glittering gold is better told in the context of environmental damage as a story of toxic processing. As Ilyas and Lee (2018) write, the toxicity and waste associated with gold mining is difficult to grasp. 'Despite the shiny, ruddy yellow colour, gold is termed "dirty" gold due to the process requirement of huge toxic chemicals and the generation of tons of wastes' (22).

Gold rush

The first recorded large gold rush started in 1693 at Minas Gerais in the then Portuguese colony of Brazil. The area was known as Ouro Preto (Black Gold) and then renamed Vila Rica (Rich Town). In the rush to the region 400,000 Portuguese and a million slaves came to mine gold. In 1848 gold was 'found' near Sacramento initiating the California Gold Rush. The yellow rock was 'found' in Eastern Australia (1851), Nevada (1859), Colorado (1875), the Witwatersrand in South Africa (1885), New Zealand and Western Australia (1892) and Western Canada (1896) (Ilyas and Lee 2018, 4).

The nineteenth century was a Goldenscene!

American museums trace the history of global migration arising from gold rushes. In the 1848 rush, around 300,000 people moved to California from the rest of the country and from overseas. The California Gold Rush found around 12 million ounces worth about 12 billion today (Richards 2016, 28). In the discourse of White Geology, 'the resulting economic boom led to increased trade and communications within the United States and around the world' (28). This obfuscates the severe impact on First Nations Peoples who were pushed off their land by the 'forty-niners', prospectors who panned for gold in streams and riverbeds. It also omits the environmental impact from the processes and emissions of gold mining and the effects of deforestation and soil and water degradation on climate change.

Gold was located by miners in 'claims' that differ with each site establishing its own rules. In Mariposa, in the foothills of the Sierra Nevada, claims were staked or 'pegged' over as much land as a miner could work, 100 square feet (nine square metres) being usual (Richards 2016, 25). Quickly a trade arose in buying and selling claims. Once easy pickings are gone at a mining site, capital and planning are required to organise mechanisation, and this requires increasing uses of energy.

The California State Mining and Mineral Museum in Mariposa houses the official California State Mineral Collection. The Collection, started in 1880 in San Francisco, has over 13,000 gems and mineral specimens, and mining artefacts. The museum is near where Kit Carson and John C. Freemont located the Mariposa Vein and built a stamp mill. The museum contains a working scale model of the mill – a mechanical crusher that demonstrates the process of crushing ore and extracting gold from quartz rock. A mining tunnel is connected to the museum, which became part of a state park in 1999. The museum presents a history of mining gold as a story of progress, 'walk through our mine tunnel and see how gold was mined in the mid-1880s, when California was a wilderness, being transformed by rapid development' (CA. Gov).

Museums in the United Kingdom trace a history of gold appearing in the West from 25 AD with the Romans whose quest for resources included gold. There are records of gold mined in the Iron Age in southern Scotland. But most gold production in Britain has come from the Dolgellau Gold Belt in North Wales. Wales has a rich heritage of gold mining with evidence of gold mines going back to Roman occupation in 74 AD. In northern Wales, the Clogau St David's mine was mined from the 1890s to early 1990s, when the high cost of mining closed it down. Exploration was underway in 2018 to locate new gold veins using recent technology. The mystique around Welsh Gold is useful to such a cause; the allure that high-grade Welsh gold is the world's most sought is supported by emotional attachment to the U.K. monarchy; since 1923 royal wedding rings have been crafted from a single gold nugget from the St David's mine.

In the Colony of Victoria from 1850 a succession of gold rushes transformed the colony. The Eureka Lead at Ballarat was an alluvial goldfield. Unfair and brutal governance led to the 1857 Eureka Stockade over the rights of miners. The story is told in the Eureka Centre in Ballarat, where the famous Eureka Flag is on loan from the Art Gallery of Ballarat. This flag, made of flimsy fabric, shows the Southern Cross, with white stars on a dark blue background. The government's imposing of harsh taxes led to agitation and the formation of the Ballarat Reform League modelled on ideals of the Chartists in Britain. Government forces attacked a group of miners at a roughly made stockade in 1854 with casualties on both sides (Wickham 2014). Reforms and miners' rights followed from the short battle and the ideals of equality and freedom continue to be commemorated at the site. The Eureka Centre promotes the nineteenth-century events on the goldfield in the current political context: 'It has been said that Eureka is the birthplace of Australian democracy. 'The effect "Eureka" had on the mining laws, of equality within society, the legislative administration and the influence upon subsequent generations of Australians has been profound' (Wickham 2014, 49).

In whatever capacity visitors absorb the messaging around the events at Eureka, the Centre does not give primary attention to the dislocation of First Nations Wadawurrung and Dja Dja Wurrung Peoples, and there is no attention to the environment impact of gold mining and processing in the region. The pedagogic function to educate visitors about Australia's democratic processes inadvertently puts into sharp focus the absence of discourse around the deep significance and dislocation of mining and industrial development in the region.

Aesthetics

Gold is universally valued across the globe, and is a marker of beauty, wealth and possession. The exploitation of earth's metals is exemplified by the value attached to gold. The visceral desire for gold relates to an abstract economy arising from gold's rarity, malleability and durability. For two centuries until 1971, global finance was underpinned by the gold standard, with paper money linked to an equivalent weight in gold. Gold 'holds its own', as the expression goes. Gold is the ultimate signifier of power. There are numerous gauges for the human character based on the effect of the yellow rock: gold digger, heart of gold. 'Gold, either as a commodity or gift, is an extremely powerful element, that, for reasons that remain largely unexplained exerts a complex fascination on human beings' (Gaitán Ammann 2006, 231). Gaitán Ammann writes that there is 'a hint of universality to human attraction for golden objects, a puzzling topic that has seldom been addressed in a satisfactory way' (239).

There is an attraction to gold that does not hold for other minerals; iron ore, zinc and copper are associated with the making of fortunes but not with the same hallucinatory desire and obsession. Gold is a story of the febrile self, expressed in the metaphor of 'gold fever'. This linguistic gesture to gold desire as an embodied experience taps into a self-interest that is considered acceptable rather than self-aggrandisement: 'To find the gold, I first had to find myself. I needed to dig deep and discover the resilience and fortitude required to overcome the solitude, remoteness, disease and violence criminals – to come out on top', writes a gold miner (Richards 2016,12).

The aesthetic of gold as a combination of beauty and power is also a recurring theme in museum displays. What seems curiously ambivalent is that the origin story of a gold discovery and ensuing rush is notable for the dreadful conditions and suffering endured by miners and their families, and enforced workers who move onto a goldfield. This is conveyed through photographs and the makeshift shelters, tools and technologies used to extract the stuff from the ground. As the story of endurance is usually told in museums, these hardships are part of the pioneering spirit of resilient individuals that led to the development of communities, citizenry and 'the nation'.

In museums, the alchemic magic of gold and its value on the stock exchange merge. Or more pragmatically, 'Museums and banks are, at a glance, two familiar places where we keep our treasures safe' (Gaitán Ammann 2006, 231). Gaitán Ammann writes of the close and problematic relationship in countries like Columbia, between a central bank and a country's art and archaeological artefacts. Commonly, and like a bank, there is a special 'vault' or 'secure' display in a museum that holds gold objects and/or replicas of gold ingots. The vault is the area where tales of heists and other gold related crimes are told. The vault at the California Mining Museum tells the story of mining by simulating the wealth of the material on display. In the Goldfields Museum in Kalgoorlie-Boulder, 'the vault' is a dramatically lit room with ingots spot lit in a showcase alongside texts and photographs depicting a robbery and grisly murder. Visitors enter the low-ceilinged room through a steel door with an obvious alarm system, a feature that heightens the identification of gold with crime.

A display of 'Gold and Greed' at the Goldfields Museum tells the 'diabolical crime' of the murder of two detectives from the Gold Stealing Detection Unit. Set up in 1907 the unit, which is still in operation, detected the theft of gold from mines. In 1926 the inspectors' remains were found dismembered in sacks at the bottom of an abandoned shaft. The vault includes a photo of the site at Miller's Shaft, the revolver used to commit the crime, a bullion bag and an ingot. The story concludes with the capture, conviction and hanging of the two murderers.

For a museum to provide a regional history of gold and its extraction, a social history of migration and place-making includes a strong measure of adventure. Gold fever initiates a rush to claim sites for prospecting in a process that is represented as a mix of madness and ingenuity. The delirium of gold fever falls into a category of its own and is resolved as progressively shaping a region and nation. The idea that the lure of gold has directed the course of human progress is conveyed in narratives that tell of acts of war, expressions of love, mass migration and technology. The devastation on the environment and the contribution of pollution to climate change are not a part of the story; rather there is an identification of mining technology with ingenuity, economic growth and social progress.

Around the world today gold is commonly mined in massive open pits. In Kalgoorlie this is the Fimiston Open Pit, known globally as the Super Pit, which runs along the edge of the city. The Pit is huge: 3.5 kilometres long, 1.5 km wide and around 700 metres deep. It is a consolidation of almost interlocking underground workings that comprised a region of one of the world's richest gold deposits known as The Golden Mile. More than 1,000 ore lodes occur within the Golden Mile. Run by KCGM (Kalgoorlie Consolidated Gold Mines) The Super Pit has steadily expanded over 25 years, blast by blast and load by load.

The Pit itself is a working museum, a popular tourist site with carefully organised trips run by company staff. Visitors are driven by a company bus down into the Pit where they are able to view at close up from various sites, the massive operations. Everything is supersized, from the enormous trucks with 66-ton buckets, to voids that appear from old mining shafts, 'big enough to swallow a substantial part of the KCGM mining equipment fleet' (Skinner 2014, 48). What I noticed on each of my tours of the Pit is the constant hum from the enormous trucks that move slowly and incessantly up and down the roads within the Pit. From sites that look down into the bottom of the Pit, the trucks are toy sized, which gives a sense of the extreme size of the gaping hole in the Earth.

Sometimes a regional museum is configured within an old mine site, or attached to a tunnel. These sites combine features of industrial heritage and social history and focus on mining inventions, individual characters and communities. They dwell on the excitement of the discovery phase. Sovereign Hill in Ballarat opened in 1970 as a 'living history museum' with three precincts: the Red Hill Gully Diggings (already discussed), Main Street and the Gold Mine. It opened around the same time as the Beamish Museum in County Durham (discussed earlier).

The story of the Ballarat goldfield continues with replicas of timber shafts and a windlass driven by a horse to reach the deeper leads of gold. With the arrival of steam power, mining companies built underground mines, including the Red Hill Mine, where visitors can go underground. Main Street is a street of shops where visitors can purchase merchandise and gold mining memorabilia. With the new museology, and in the spirit of inclusive social history, the dispossession of First Nations Peoples and the plight of woman and migrant labour are given space but these function within the heteronormative narrative of fortitude, endurance and nation building.

Golden alienation

Gaitán Ammann (2006) explores the politics of the Gold Museum in Bogota. He finds that the purpose of the museum to safeguard pre-Hispanic gold has been achieved 'at the price of a striking essentialization of indigenous social life in the past, and a curious disregard for the commodity value of ancient goldwork in the present' (230). Popular concepts of political and economic development, such as those told at the museums, 'have contributed to marginalize present-day indigenous groups in Columbian society'. In a frank appraisal of the damage caused by the museum's backing by the Central Bank of Columbia, he writes: 'The staggering rupture between past and present that the Gold Museum displays is a complex phenomenon I term a *golden alienation*' (230).

The Gold Museum's collection was initiated by acquisitions purchased in the 1940s by Columbia's Central Bank, who 'openly purchased hundreds of gold artifacts from private collectors eager to exchange them for readily usable bank notes' (Gaitán Ammann 2006, 233). The museum started as a private venture of the Bank with visitors invited to a display lounge by appointment. In 1959 the first exhibition was held for a limited public. 'Enrobed in heavy curtains and draperies, gold masterpieces were displayed in a secretive atmosphere that did little to stimulate the interpretative capabilities of the museum's visitors' (234). In 1968 the museum opened in a new minimalist building. Like other gold museums, this contained a security vault or Golden Room that recreated the security setting of a bank with a light and music display that focused on the myth of El Dorado. The museum also focused on interpretations of pre-Hispanic metallurgy that 'enriched the understanding of ancient cognitive systems in northern South America' (235). The museum moved into another new building in 2003 that doubled the permanent exhibition area.

Gaitán Ammann's analysis of the Gold Museum admits the museum is rightly celebrated for its remarkable pre-Hispanic goldwork; however he qualifies this by the way the museum perpetuates 'a dazzling and unsettling testimony of humankind's obsession with the materiality of gold' (237). He finds it problematic that the museum highlights the technological sophistication of the goldwork but Indigenous people remain stigmatised, uneconomic second-rate citizens in Columbian society (238). Michael Taussig (2004) also argues the link between African slavery and the socio-economic relevance of gold mining is overlooked in the aestheticisation of Indigenous goldwork (Ammann 2006, 245).

In a recent memoir, a gold miner observes that 'mining is messy, some of it is destructive and at times it is downright lethal'; however the destruction is glossed over by the fact that 'the industry also supports a vast web of otherwise impoverished and marginalised people' (Richards 2016, 11). Here we find the argument that capital is community that without capitalism modern society would not exist. Gold mining signifies the creation of order out of chaos, with the wealth of nations tied to its industry and economic growth. This alignment reinforces the division between nature and humans that is necessary to qualify by telling the truth about the destruction of mineral extraction and toxic processing. As museums confront the climate emergency, it will be a challenge to reconcile the powerful narrative of economic growth with the reality of environmental damage, and species habitat destruction. The carbon emissions generated by gold mining are huge. Given this, a museum can confront growth ideology by going inside, by considering the perspective from the point of view of gold.

It is pertinent in the story of gold processing that once it is 'recovered' from the treated slurry of which is a crushed component, it is permanent and incorruptible. Imagine the story it can tell! Gold will remain the same, long after humans have gone. There is a gold geopoetry that elides the obsession attached to the mineral. It is a poetry that exposes the paradox of basing the brief and mortal life of a human on the stability and durability of gold.

Unlike the pragmatic link of iron and fossil carbons with an iron age and Industrial Revolution, gold signifies adventure, crime and madness. Gold is associated with historical acts of horrendous violence. The Spanish conquest of the Inca Empire by Francisco Pizarro arose from gold fever that destroyed the kingdom of Emperor Atahualpa. Pizarro ambushed and captured Atahualpa at Cajamarca, a small city in the Andes in northern Peru. He slaughtered thousands of unarmed soldiers and, in a museum known today as the Ransom Room, ordered his people to fill the room with gold. 'Sun disks, ritual objects, and gold plates that covered the inner walls of temples were brought [to the room] and the Spanish melted them into ingots' (Lingis 2018, 57). The emperor was strangled, and Pizarro and 168 Spaniards marched on to Qosqo and the conquest of Peru.

The brutal violence attached to gold carries into fiction, such as the murder of Viserys Tagaryen in the popular HBO fantasy series of *Game of Thrones*. The sadistic greedy Viserys, desperate to be crowned, dies when molten gold is poured over his head. A hardly better death than the recorded Roman act of execution by pouring molten gold down a victim's throat (Science Alert 2014). Ill-fated endings and abject intimacies linger around gold. The ancient Phocylides held a clear view on its lure:

> Gold and silver are injurious to mortals; gold is the source of crime, the plague of life, and the ruin of all things. Would that thou were not such an attractive scourge! Because of thee arise robberies, homicides, warfare, brothers are maddened against brothers, and children against parents.
> (Agricola [1556] 1950, 8)

Alphonso Lingis reports on contemporary conflict in Peru over the proposed expansion of the Yanacocha gold mine in the mountains above the ancient Incan city of Cajamarca. The open pit mine covers 535 square miles and is well known for its toxic pollutants: 'The people have found that the rivers and streams and their wells are polluted with mercury. Their animals have died;

children have gotten sick' (Lingis 2018, 61). The company that runs the mine is owned by Newmont Mining Corporation from Colorado, the Peruvian Mining company Buenadventura and the World Bank's International Finance Corporation (61). Lingis (2018) writes:

> In 2010 Newmont Mining Corporation and Buenaventura launched the $4.8 billion Conga project to vastly expand the mining operations above Cajamarca. For water the company will use its cyanide extraction process, it would harness four mountain lakes that provide water to a hundred farming communities and the city of Cajamarca. To replace the lakes, the company proposes to create four reservoirs that would be filled by rainwater. The local population has repeatedly assembled in protest. Five people have been killed and some fifty wounded.
>
> (62)

There is much gold deep in the Earth and a journey to the core would reach these layers; currently however, if all the gold mined 'formed into one solid cube, it would fit on board a single oil tanker' (Lingis 2018, 57). Gold has always been associated with alchemical and healing properties. The chemistry of gold originates from exploding stars (as ancient alchemists claimed). Clouds of gas and dust from these explosions coalesced into stellar nurseries from which earth's solar system was born. Gold was flung into space and some fell to earth in meteor showers; as earth formed, molten iron sank to the centre taking with it the gold and other precious metals (Seeger et al. 1964; Willbold et al. 2011).

Gold's superpowers continue to be framed as an agency that is under human control: as science advances it is acknowledged to be a transporter around living bodies; 'you can stick a bunch of things on a gold nanoparticle and get it to have very controlled behaviour' (Weintraub 2013, 1). But a geopoetry of gold would expose different stories that are relevant in a climate emergency, and that assemble the tragedy of gold fever on First Nations Peoples, with the terrible burden on the environment.

After neutrality
The relevant museum

'It no longer seems rational to assume that humanity, encountering an existential threat, will behave rationally', writes journalist Nathanial Rich (cited in Huntley 2020, 17). The nebulous forces of fossil fuel driven capital continue to drive the threat to life of the warming globe. Tim Flannery (2020), a scientist closely involved in climate change politics, documents the role of Australia, an affluent developed country, in disrupting international attempts at reducing carbon emissions (147). Flannery is one among many scientists, journalists and activists trying to drive action by exposing the political self-interest behind the climate change debate. As a strong liberal democracy, Australia has not dealt rationally with the threat of climate emergency by putting a brake on emissions. From 2013 it has been a wrecker of global climate action while also becoming 'the world's largest exporter of two of the world's three fossil fuels – coal and gas' (Flannery 2020, 30). The country can be held up as an exemplar of The Great Dithering.

In democracies like Australia there is a level of open concern and debate about the short-term focus and duplicity of politics, a debate that does not happen in countries where speech is stifled. But the example of Australia indicates the complex governance required to deal with the climate emergency. Like other developed nations in the global north the country suffers from excessive consumerism or 'affluenza' (Hamilton and Denniss 2005), with successive governments subservient to Big Mining and the mantra of economic growth.

The role of museums is significant in confronting these concerns and the wicked problem of economic growth by the action of taking a united stance on climate change and global warming. Museums cannot ignore the politics of climate by claiming the privilege of being neutral public institutions. The politics of climate in the 2000s are too thoroughly exposed, for example, in the Australian context, Clive Hamilton (2007) observes:

> ... climate change policy in Canberra has for years been determined by a small group of lobbyists who happily describe themselves as the 'greenhouse mafia'. This cabal consists of the executive directors of a handful of industry associations in the coal, oil, cement, aluminium, mining and

DOI: 10.4324/9780367741945-11

electricity industries. Almost all of these industry lobbyists have been plucked from the senior ranks of the Australian Public Service, where they wrote briefs and Cabinet submissions and advised ministers on energy policy.

(4)

Fossil fuel capitalism will not support a rational position on climate as this is not in its interest; its eye will be on using any disruption to its own advantage. And so it is that museums must come to the fore and expose the politics of greenwashing that constrains change and enables business as usual. Museums that continue to use the cover of neutrality will increasingly be sites of misinformation.

The paradigm shift is to acknowledge *homo sapiens* as one species among many facing extinction. Inorganic actants, physical systems and life are assemblages; they are not separate and distinct. This observation is not new; it has been acknowledged in the modern context since thinkers and artists critiqued the effects on the environment of the Industrial Revolution. And it has always been the insight of First Nations Peoples. What is novel is to bring the notion into the consciousness of institutions of authority so that geo-inclusivity becomes common sense. With recognition we are a part of the material world, it is harder to exploit and degrade that world.

The notion that nature is something 'out there' to be tamed, quelled or commercialised by large corporations is foolhardy. That this acknowledgement should still be a radical proposition for institutions is telling of contemporaneity as a culture of excess and inequity. Rejecting the nature and culture divide has to be a united stance with museums at the vanguard through policies, practices and protests that communicate proper entanglements, assemblages and compositions of the material world. These entanglements restore an ethics of care that rejects the status of the inanimate as a resource to dynamite, crush, burn, dissolve, mash, flatten, seal, stabilise and contain.

The turn to alterity as an intellectual rebellion and non-violent coup in museums will be based on resisting the increasingly anti-critical stance of institutions. It is long overdue for the corporatisation of cultural institutions to be acknowledged for what it is, and for institutional rivalries to be put aside to confront the conceptual, ideological and existential challenges facing relations with earth's systems.

A focus on the maintenance and care of systems will mark a difference. While institutional focus on innovation is important (notably the science of medicine), it is not the primary goal for museums. It is not that care and maintenance have disappeared from museums but these essentials are masked or negated by the neoliberal axiom to deregulate and privatise. In our late liberal era, there remain fugitive spaces within museums that can ignite museological practices that acknowledge the Earth and its systems require urgent attention.

On fire

Using the 2019/20 megafires across southern and eastern Australia as a climate tipping point, Flannery (2020) writes the inaction of governments is no longer acceptable (11). He has become aware that 'those wanting action wrongly think that presenting facts will lead to a solution' (11). After 20 years of watching misleading acts of self-interested parties, it is time to take action. It is here that the governance and leadership of museums become imperative, that they utilise the power and knowledge relationship that is their legacy. Museums are institutions with the public trust to speak truth to power. The museum sector must end the moderate approach to the science of climate change and clearly proclaim a climate emergency. It can use its collections and practices in all sorts of imaginative ways to stage the reality of tipping points.

Tipping points are used by climate scientists to give certainty to phenomena that are usually assumed to be ongoing. We assume, for example, that the earth will carry on with the occasional volcano or other geological event. But this is not what the science modelling says. Key tipping points are the irreversible melting of the Greenland and/or Western Antarctic ice sheet, the slowing down of the North Atlantic deep ocean circulation, release of gas hydrates and die back of the Amazonian rainforest (Maslin 2014, 98).

Each country is avoiding or contending with their own tipping points; however all events are interconnected. In Western Europe, California and Australia there is an increase in extreme fire events or megafires. A year following the Australian 2019/20 megafire, while I write in Perth, there is a small but uncontrolled fire on the outskirts of the city. The temperature outside is another day of 38°C (100 F) and a tropical low from the north threatens to bring strong winds that will fan the blaze. The sky is an ominous yellow and fine ash is falling in my garden, a weird burnt snow. We can smell the fire, which is burning 40 km away. I am safe (surely) as my house is in the inner city, but have local public radio playing, having become familiar with the broadcasting of regular updates on where a fire is heading and who should evacuate or be prepared to stay and fight.

Australia is not alone in confronting disaster with rising temperatures. California confronts a similar wildfire situation and with it an element of normalisation that has come with the regularity of catastrophic fires. Writing on the 2019 Kincade Fire in Sonoma Country, Meehan Crist (2019) observes that people find ways to make 'increasingly unfamiliar and unsettling circumstances bearable', for example, 'rich Californians are hiring their own private firefighters, at a cost of around $2500 a day. She observes the near absence in coverage of the Kincade Fire being linked to climate change or development.

Fire is an increasing impact and with fire seasons in the Southern and Northern hemispheres now overlapping, North America and Australia share firefighters. Medical doctors warn that the extended fire events, droughts and floods in Australia are pushing the demands of health services, with increasing deaths from heat stroke (ABC RN 3 November 2019). In Canada in 2021, unprecedented temperatures led to deaths and fires. Having had my own experience of such a climate event makes Kim Stanley Robinson's (2020) fictionalisation of the Great Indian Heatwave, an event that killed 20 million, a harrowing prediction of what is to come unless united action is taken now.

Hubris

Geological entanglement challenges the hubris of excessive production and consumption that enables gross mismanagement of the earth. Meehan Crist (2019) confronts the hubris of advertising following the Kincade Fire.

> One morning nearly a week after the Kincade Fire first started burning, I went online for the overnight news and was met with a sleek commercial for Lexus cars: 'Is it possible to outsmart fate?' A slim, handsome man with carefully rumpled beard smiled at me from the side of his face.

Such blatant hubris requires attention to the narratives being generated by petrocapital and its products and to understand that the extraction industries continue to hold power over cultural norms. It displays what cultural theorist Mark Fisher (2009) calls a business ontology in his analysis of capital realism. 'Climate change and the threat of resource-depletion are not being repressed so much as incorporated into advertising and marketing' (18). Capital realism mocks the fact that human and inorganic elements are inseparable.

Instead of the earth being silenced, marginalised and/or exploited in systems of endless growth, the critical stance is to question the way these actants feature in modern institutions.

It is not the time to dispense or deaccession museum collections, but to get uncomfortable with objects and collections. To reframe and repurpose collections so they confront the museum's legacy of engagement with the hubristic discourses of western humanism. As a city, town or region is impacted by the disruptions of changing climate, this will suggest the type of encounter for museums to generate through their collections. The focus turns to engaging with the reality confronting a particular place in the present.

Staying with the topic of fire, given that drought and fire are increasing inclusions in the lives of many communities, the abject has a role in consciousness raising. Like many issues considered in this book, there is already a museum relation with abject objects. The fascination with the other that abject objects provoke is a complex area of psychology and affect. It is a matter of bringing this area into the realm of climate change, mitigation and adaptation.

Museum Manager Christine Hansen (2018) sought to understand encounters with abjection in the wake of the destruction of the 2009 Victorian Black Saturday fires. She observes, 'the community was incredulous, traumatized, and, perhaps counterintuitively, enchanted' (1). Hansen

(2017) recognises a reality to events that are too massive to be assimilated as she considers the effect of the collection of burnt objects from the fires. In the malformed objects, she finds a vibrant materiality of entanglement. Oxidised materials were melted beyond recognition into monstrous forms; a 'car windscreen melted into a cascade of hardened dripping glass: the things of our lives dissolved and reformed in the monstrous kiln of a firestorm' (Hansen 2017, 129). Instead of Amy Balkin's *People's Archive of Sinking and Melting*, this is an Archive of Burning, and is similarly poignant and relevant to the lost and disappeared environments the artefacts represent. Such displays are profoundly relevant to communities who have experienced the trauma associated with drought, fire and flooding, as they confront, acknowledge and warn against loss rather than hiding it from view.

In museums, devastated objects acquire everyday meaning that affects the object's abject quality. I am reminded of twisted bicycles and burnt bricks in the Hiroshima Museum (Baker 2017b, 112) shown in the film *Hiroshima Mon Amour* (Alain Resnais 1959) as well as the banality of tragedy reflected in Balkin's project. Hansen seeks to understand such object allure when she describes the feeling of viewing burnt artefacts as that of 'fire on flesh'. The lure of these artefacts is not an object transformed by wildfire but a sense of felt horror: 'What we are collecting is not the material evidence of a chemical phenomenon, but of our horror'. Here is an instance where tragedy becomes 'an affective fascination with destruction' (Hansen 2017, 131). What remains in the aftermath of fire is glass, metal, clay, concrete, rock and occasionally bone; the remainder of the material world is obliterated (2017, 128). Hansen (2018) writes of the fires:

> No one directly experienced the heart of the firestorm and lived; if you were in its path you perished. From afar, however, it was riveting, a cyclone of flame spinning its own pyrocumulus tower stretching fifteen kilometres above the earth, filling the air with ash and turning the sun red. It was an environmental drama of an order rarely experienced. It its scale, it was awesome.
>
> (227)

It is my belief that museum visitors are ready to confront the psychological depth of encounters with climate disruption and through this their complicity with fossil capital and outdated consumer behaviour. They are experts in being duped by fossil capital, and a role of museums is to articulate modes of misinformation. Visitors to museums are ready to be exposed to the terrible lure of objects as this provides a shelter from the technologies of deception that bombard contemporary consumer culture. Enough people understand they are the target of greenwashing, and that the slow shifts of policy by governments appease constituents and do not represent leadership toward genuine change. Communities of people know they are being duped even while participating in the system that dupes them. We consume the products churned out by Amazon, Google, Apple and other massive enterprises while at the same time realising that the consumer way of life is coming to an end.

An end can be a beginning. Jeff Sparrow (2021) and other critics of market economics find hope that change is underway.

> Precisely because we face a crisis existential in its implications, minimal programs fail to inspire. We all know, deep in our guts, that we won't get of this without massive change. The cautious, "sensible" proposals put forward by politicians don't inspire anyone. Perversely, they seem utopian: a fantasy of moderation in which no one really believes.
>
> (199)

This book has not been a comprehensive survey of museums and their management around climate awareness (although there is some of this); rather it has sought to draw on theoretical and creative approaches to climate by thinking via the geological what museums can become. This provides a framing for alternative stories to the anthropocentricism that has a tenacious purchase in museums, including climate programs and exhibits. Institutions that avoid the climate emergency are not only irrelevant; they do harm. Climate initiatives in museums that sidestep the insights conveyed by thinkers, scholars and artists who expose relations between climate, colonialism and capitalism through concepts such as Capitaloscene, Plantationscene and Chthulucene are clouding the truth and contributing to misinformation.

Lithic entanglements are everywhere yet diminished by an antiquated humanism, by a fossilised humanism. Better forwarded is an anti-humanist or inhuman code of conduct that contests the belief systems that enable material exploitation. The exploitation is dangerously justified through modern epistemologies and ontologies that classify geological formations and separate them from the affairs of humans. In this ontology rocks, minerals and soils are world less. They have no voice or agency. The museum can change the status quo by providing a lithic gaze onto the human. To accentuate knowledges other than western colonialism, which is now a neoliberal numbness. Here I position museums as makers of odd kin, a geontological mud or composition of thoughts and feelings able to challenge dominant paradigms. These de-centred ways of knowing are tools with which to confront the disingenuous project of fossil fuel capitalism.

Museum practices have developed in tandem with dualistic thinking, which is problematic in approaching the object of climate change as 'content tends to disregard the non-human (biosphere, ice, coal, greenhouse gases), therefore presenting the non-human merely as a passive entity and subject for intervention, rather than as actants in their own right' (Cameron 2015, 357). Cameron's museum that 'conceives of institutions as fluid assemblages composed of heterogeneous collectives of human and non-human actants' (ibid 356) is exemplary of scholarship on museum practice that offers a vital alternative. The insights of contemporary art that acknowledges the damaged earth and climate emergency also provide ways to address the current moment. Heather Davis and Etienne Turpin (2015) position art as a 'non-moral form of address that offers a range of discursive, visual and sensual strategies that are not confined by the regimes of scientific objectivity, political moralism, or psychological depression' (4).

All public museums and sites – natural history, living history, science, art, industrial – are tasked to find new ways to organise knowledge and to activate this knowledge. With such a remit, and with their high credibility factor, museums can challenge what is normal and common sense. For how is normal to be measured as melting ice, rising seas, drought, floods and fires bring more and more species into visceral proximity with fear, grief and extinction? Museums must step into this terrifying space, rise to the mark, take risks. The remainder of the twenty-first century is no time for cosy directorships and light-weight leadership in cultural institutions.

Climate change and the museum's future as that of a sage philosopher are the resolve of this study. But the museum is not an academic philosopher, who is not required to make abstract ideas lucid for a non-expert audience. The museum is a medicine for academic self-interest at one end of the spectrum and the takeover of misinformation at the other. The climate is not out there far away; it is what we live and think with; the weather is us. It is the milieu we move through and through which we feel and think.

Responding to climate as a force not separate from being human, from our bodies, requires the unsettlement of received thinking about what it is to be human. This is a powerful force for new relevance in museums, as it takes the institution beyond the instrumental commodification of culture and nature. This does not mean throwing out everything modern and returning to a pre-Enlightened age. Rather it seeks to move past the categories that have set up what it means to be modern; it sits with Bruno Latour's claim that 'we' have never been modern (Latour 1993). We are a species; it is how we think about the materiality of being that is the point. In this, the museum can repurpose the knowledge it previously supported. The process of repurposing knowledge is a method of communicating new ways of becoming, new subjective positions, a new subject. An epistemological revolution can happen in and through the museum that adopts a new ontology of inclusion.

There is significant civic value in this reimagining as it exposes the divisive inequalities wrought during the neoliberal phase of western humanism. The philosophical task for curators and museum workers is enormous – to frame collections and practices differently, to develop programmes that present the focus on innovation as over-zealous, to carefully consider the various claims of posthumanism, to query trans-humanist desire for an ideal human, to present the danger of the world continuing as a Technosphere run on massive consumption of energy. These and other substantive issues communicate a climate emergency. Museums can present a revaluation of the structures that are the site for real transformation.

Hope

There is much for museums to expose. Even theories that appear to support radical change have difficulties that need to be considered. With posthumanism(s) emerging as a dominant critical discourse on environmental and ecological issues, 'rocks' slide back to one thing among other agents; there is little focus on including the alterity of the inorganic. This is the realm for museums to engage through inventing neologisms and new strange sounding concepts. It is what post*humous* approaches enable; kinships that make assemblages of humans, minerals and physical systems that operate at the threshold of life and nonlife.

In moving toward what thinking must become, it is also apparent that we have become numb to information and data. Claire Colebrook notes, 'If climate change politics has taught us anything to date … it is that information and data … has not yielded any affective response'. Mike Hulme and climate scientists admit the same. How can the climate emergency as here-now be affected so that action commensurate to the event is taken? For Colebrook (2016) concepts such as Anthropocene are indicative of a heightened human-centrism that avoids confronting reality.

> … the critical impact of notions of anthropogenic climate change and the Anthropocene has been enlivening, rather than devastating; precisely when we ought to be confronted with 'civilization' as a trajectory of wreckage, we become all too focused on surviving.
>
> (114)

To ask whether we ought to survive is perhaps the final controversy. But what this question forces us to ask is why is this question so difficult. Colebrook considers that as long as life presents itself as that which must be sustained the difficult question of how human life can justify itself is avoided. We cannot ask how we might survive 'because the "we" of the question is at once that which has defined life *and* that which is essential hurling towards its own extinction' (Colebrook 2014, 204).

Again, this is difficult territory that museums can lucidly frame.

There is attention in museum research to the catastrophic impacts of human activity on living species, with the spectre of a sixth extinction hovering over the biological and zoological sciences. The Directors and CEOs of Australian natural history museums issued a statement following the country's catastrophic 2019/20 fires:

> We now recognise human-induced climate change, alongside land clearing and habitat use, as the over-arching issue affecting Australia's unique wildlife as evidenced by more intense bushfires, drought, floods and the impact of warming oceans on the Great Barrier Reef and other marine environments.
>
> (Australian Museum 2020)

The Museums Association (MA) in the United Kingdom have initiated a Museums for Climate Justice Campaign with the purpose 'to support museums in tackling the climate and ecological crisis' (MA 2022). The campaign understands that climate is an ethical, social, environmental and political issue rather than one that is purely scientific and physical. It aims to assist museums across the United Kingdom to be bold. As well as advocacy, MA intend to reduce their own climate impact, and to embed climate activism across all their activities through taking an intersectional approach that has a focus on climate and social justice.

Through its legacy as an institution of authority when it comes to world-making – museums can estrange the power relations of the White Geology through presenting encounters with its own anomalies. A geo-inclusive museum articulates the physical, emotional and psychological destruction caused by economic, political and cultural systems unable to cope with what thinking about the earth must become. They are sites to critique the arthritic rhetoric that only advances the choice of market capitalism and growth economics without taking into perspective the damage and danger of these projects. Thinking differently about production and consumption is an adaptation to climate change.

There is an imperative to gather intellectually rigorous approaches to more-than-human worlds and in doing so to actively acknowledge the intrinsic value of First Nations Peoples ontologies. Indigenous cultures understand that listening to granular shifts in land and Country is proper common sense. In response to extreme climate events and with disillusionment in the modern subject this knowledge is increasingly respected as hope for the future.

Regardless of Big Tech and the value it accords to perfectibility achieved through technology, the destruction of the world it supports is an effect of climate disruption. The climate-becoming-weather is forcing awareness that more-than-human, geological actors control the future of life on the planet. The dismantling of the modern subject that has been shaped by that subject's exploitation of the planet is underway, and with this there is a backlash. There is always a backlash when the power of an elite is threatened. Again, this can be anticipated and exposed by bold museums.

Attending to geology, and a geological way of seeing, can be traced from antiquity and medieval thinking into creation stories that evolve into the modern stratigraphic ordering of the earth. The museum is a metaphor and physical enactment of the ordering of time through objects, and a potent exemplar of enlightenment reasoning that is with us today. In this, museums are a physical staging of empiricism. Ideas are given presence through objects that are artefacts of reasoning across thousands of collections that collectively represent history and human life as a linear progression. What linear progressions assume is a clean break between then and now, between a modern world of enlightened thought and everything that went before. The museum is the material presence, witness and judge of this clean break, where no neat break exists. It is this world that needs to end.

Crutzen proposes the Anthropocene begins with the steam engine and the global impact of emissions on Earth's physical systems. The new geological time as a stratigraphical measure is an effect of industrial modernity. It is telling that the modern museum is of the same phenomena, an artefact of the industrial age. A stratotype that records enlightenment thinking as a separate time period from the pre-modern era. The eighteenth century, when industrial pollutants began to foul the atmosphere, is correlated to the age of reason, and to the collectors who identified as men of the enlightenment.

The totalising theories represented in museums remain the dominant method of visualising human and earth relations. New stories are needed to communicate that these narratives are limited by putting the human at the centre, and limiting rocks to elemental time pieces awaiting discovery and extraction. New ways of apprehending soils and minerals are an intervention showing that the museum's stage crafting of classificatory orders masks more complex epistemologies and ontologies. Like science, the museum is never a detached observer. It has finely tuned the practice of reasoning to make abject or untrue that which does not appear reasonable. It is from this starting point that the modern humanist gaze needs to be resisted. With the death of this world, the earth can then tell stories of the museum. 'It is the earth … that an antihumanist alternative needs to start from' (Colebrook and Weinstein 2017, 228).

References

Adams, Clive and Daro Montag. 2015. *Soil Culture: Bringing the Arts Down to Earth*. Devon: Centre for Contemporary Art and the Natural World.
Agricola, Georgius. (1556) 1950. *De Re Metallica*. Translated by Herbert Clark Hoover and Lou Henry Hoover. New York: Dover Publications.
Albrecht, Glenn. 2015. 'Solastalgia: A New Concept in Health and Identity'. *PAN: Philosophy, Activism, Nature*, 3 (3), 41–55.
Albrecht, Glenn, Gina-Maree Sartore, Linda Connor, Nick Higginbotham, Sonia Freeman, Brian Kelly, Helen Stain, Anne Tonna and Georgia Pollard. 2007. 'Solastalgia: The Distress Caused by Environmental Change'. *The Royal Australian and New Zealand College of Psychiatrists*, 1, S95–S98. DOI: 10.1080/10398560701701288
Almarcegui, Lara and Gerd Wessolek. 2019. 'Wastelands'. In *Field to Palette: Dialogues on Soil and Art in the Anthropocene*, edited by Alexandra Toland, Jay Stratton Noller and Gerd Wessolek, 581–590. Boca Raton: CRC.
Anderson, Nicole. 2017. 'Pre-and Posthuman Animals: The Limits and Possibilities of Animal-Human Relations'. In *Posthumous Life: Theorizing Beyond the Posthuman*, edited by Jami Weinstein and Claire Colebrook, 17–42. New York: Columbia University Press.
Armiero, Marco. 2018. 'Sabotaging the Anthropocene Or, in Praise of Mutiny'. In *Future Remains: A Cabinet of Curiosities for the Anthropocene*, edited by Gregg Mitman, Marco Armiero and Robert S. Emmett, 129–139. Chicago and London: The University of Chicago Press.
Australian Museum. 2020. Statement on 2019/2020 bushfires. https://australian.museum/learn/climate-change/2019-2020-bushfire-statement/
Autin, Whitney L. and John M. Holbrook. 2012. 'Is the Anthropocene an Issue of Stratigraphy or Pop Culture?' *GSA Today*, 22 (7). DOI: 10.1130/G153GW.1
Aw, Titah. 2020. 'A Marriage Ceremony for Rocks Is a Hotspot for Creatives in This Indonesian Village'. Translated by Jade Poa. https://vice.com/en_au/article. 20 January.
Baker, Janice. 2013. 'Defacing the Acquisitions: A Museal-Analysis of Serial Killing Horror in Cinema'. In *Murders and Acquisitions: Representations of the Serial Killer in Popular Culture*, edited by Alzena MacDonald, 163–180. New York, London: Bloomsbury.
Baker, Janice. 2017a. 'Metal Fictions'. In *CTheory.net Theorizing 21C*. 16/5/2017. https://journals.uvic.ca/index.php/ctheory/article/view/16858/7179

References

Baker, Janice. 2017b. *Sentient Relics: Museums and Cinematic Affect*. London and New York: Routledge.

Baker, Janice. 2022. 'From-the-Hip: Rocks and Critical Heritage Ecology in the Western Australian Pilbara'. In *Heritage Ecologies*, edited by Torgeir Rinke Bangstad and Þóra Pétursdóttir, 165–182. London and New York: Routledge.

Baker, Matt and John Gordon. 2013. 'Unconformities, Schisms and Sutures; Geology and the Art of Mythology in Scotland'. In *Making the Geologic Now: Responses to Material Conditions of Contemporary Life*, edited by Elizabeth Ellsworth and Jamie Kruse, 163–169. New York: Punctum Books.

Barad, Karen. 2017. 'No Small Matter: Mushroom Clouds, Ecologies of Nothingness, and Strange Topologies of Spacetimemattering'. In *Arts of Living on a Damaged Planet*, edited by Anna Tsing, Heather Swanson, Elaine Gan and Nils Bubandt, G103–G120. Minneapolis and London: University of Minnesota Press.

Bateman, Jessica. 2021. BBC Future Planet. 'The end of the world's capital of brown coal'. www.bbc.com/future/article/20210419-the-end-of-the-worlds-capital-of-brown-coal. Accessed 23 April 2021.

Beckman, Frida. 2017. 'Posthumanism and Narrativity: Beginning Again with Arendt, Derrida and Deleuze'. In *Posthumous Life: Theorizing Beyond the Posthuman*, edited by Jami Weinstein and Claire Colebrook, 43–64. New York: Columbia University Press.

Belisle, Brooke. 2013. 'Artifacts: Trevor Paglin's Frontier Photography'. In *Making the Geologic Now: Responses to Material Conditions of Contemporary Life*, edited by Elizabeth Ellsworth and Jamie Kruse, 145–149. New York: Punctum Books.

Bennett, Jane. 2010. *Vibrant Matter: A Political Ecology of Things*. Durham and London: Duke University Press.

Bennett, Jane. 2013. 'Earthling, Now and Forever?' In *Making the Geologic Now: Responses to Material Conditions of Contemporary Life*, edited by Elizabeth Ellsworth and Jamie Kruse, 244–246. New York: Punctum Books.

Bennett, Tony. 2004. *Pasts beyond Memory: Evolution, Museums, Colonialism*. London and New York: Routledge.

Beresford, Quentin. 2018. *Adani and the War over Coal*. Sydney: New South Publishing.

Boulton, Elizabeth. 2016. 'Climate Change as a Hyperobject: A Critical Review of Timothy Morton's Reframing Narrative'. *WIREs Climate Change 2016*, 7, 772–785. DOI: 10.1002/wcc.410

Bowman Sculpture. n.d. *Emily Young*. www.Bowman Sculpture.com

Boxell, Mark and Will Wright. 2017. 'Postwar Play and Petroleum: Tourism and Energy Abundance in Rocky Mountain National Park'. *Journal of Tourism History*, 9 (2–3), 119–138.

Braidotti, Rosi. 2011. *Nomadic Theory: The Portable Rosi Braidotti*. New York: Columbia University Press.

Braidotti, Rosi. 2013. *The Posthuman*. Cambridge, UK: Polity.

Bronstein, Michaela. 2019. 'Taking the Future into Account: Today's Novels for Tomorrow's Readers'. *PMLA Publications of the Modern Language Association of America*, 134 (1), 121–136.

Brown, Kathryn. 2020. *The Routledge Companion to Digital Humanities and Art History*. Milton: Taylor & Francis.

Cameron, Fiona. 2015. 'The Liquid Museum New Institutional Ontologies for a Complex, Uncertain World'. In *Museum Theory*, edited by Andrea Witcomb and

Kylie Message. *The International Handbook of Museums Studies*. General editors, Sharon Macdonald and Helen Rees Leahy, 345–361. Wiley Blackwell.

Cameron, Fiona R. 2017a. 'Ecologizing Experimentations: A Method and Manifesto for Composing a Post-humanist Museum'. In *Climate Change and Museum Futures*, edited by Fiona R. Cameron and Brett Neilson, 16–33. New York and London: Routledge.

Cameron, Fiona R. 2017b. 'We Are on Nature's Side? Experimental Work in Rewriting Narratives of Climate Change for Museum Exhibitions'. In *Climate Change and Museum Futures*, edited by Fiona R. Cameron and Brett Neilson, 51–77. New York and London: Routledge.

Cameron, Fiona R. and Brett Neilson, eds. 2017. *Climate Change and Museum Futures*. New York and London: Routledge.

Carnegie Museum of Natural History. Anthropocene Studies. https://carnegiemnh.org/research/anthropocene/. Accessed 21 May 2021.

Center for Land Use Interpretation. 2013. 'Terminal Atomic: Technogeomorphological Mounds'. In *Making the Geologic Now: Responses to Material Conditions of Contemporary Life*, edited by Elizabeth Ellsworth and Jamie Kruse, 238–241. New York: Punctum Books.

Chakrabarty, Dipesh. 2009. 'The Climate of History: Four Theses'. *Critical Inquiry*, 35 (2), 197–222.

Chakrabarty, Dipesh. 2019. 'Museums between Globalisation and the Anthropocene'. *Museum International*, 71 (1–2), 12–19. DOI: 10.1080/13500775.2019.1638022

Chandler, Jo. 2021. The Science Show. *ABC Radio National*. 17 April www.abc.net.au/radionational/programs/scienceshow/climate-change-is-f*ing-terrifying.-has-the-media-failed-in-t/13305506. Accessed 19 April 2021.

Chong, Derrick. 2012. 'Institutions Trust Institutions: Critiques by Artists of the BP/Tate Partnership'. *Journal of Marketing*, 33 (2). DOI: 10.1177/0276146712470458

Clark, Nigel. 2014. 'Geo-politics and the Disaster of the Anthropocene'. *The Sociological Review*, 62 (S1), 19–37.

Cohen, Jeffrey Jerome. 2015. *Stone: An Ecology of the Inhuman*. Minneapolis and London: Minnesota Press.

Colebrook, Claire. 2014. *Death of the Posthuman: Essays on Extinction, Vol. 1*. Ann Arbor: Open Humanities Press with Michigan Publishing.

Colebrook, Claire. 2016. 'What is the Anthropos-Political?' In *Twilight of the Anthropocene Idols*, edited by Tom Cohen, Claire Colebrook and J. Hillis Miller, 81–125. London: Open Humanities Press.

Colebrook, Claire and Jami Weinstein. 2017. 'Preface: Postscript on the Posthuman'. In *Posthumous Life: Theorizing Beyond the Posthuman*, edited by Jami Weinstein and Claire Colebrook ix–xxix. New York: Columbia University Press.

Cook, Jill. 2003. 'The Nature of the Earth and the Fossil Debate'. In *Enlightenment Discovering the World in the Eighteenth Century*, edited by Kim Sloan, 92–99. London: The British Museum Press.

Crist, Meehan. 2019. 'California Burns' Diary. *London Review of Books*, Vol. 41: 22. 21 November.

Cross, Gary and John K. Walton. 2005. *The Playful Crowd: Pleasure Places in the Twentieth Century*. New York: Columbia University Press.

Crutzen, Paul J. and Eugene F. Stoermer. 2000. 'The "Anthropocene"'. *IGBP Newsletter*, 41 (May), 17–18.

Culp, Andrew. 2016. *Dark Deleuze*. Minneapolis: University of Minnesota Press.

Cuno, James. 2011. *Museums Matter: In Praise of the Encyclopedic Museum*. Chicago and London: The University of Chicago Press.

Davies, William. 2015. *The Happiness Industry*. London: Verso.

Davis, Heather. 2015a. 'Life & Death in the Anthropocene: A Short History of Plastic. In *Art in the Anthropocene: Encounters Among Aesthetics, Politics, Environments and Epistemologies*, edited by Heather Davis and Etienne Turpin, 347–358. London: Open Humanities Press.

Davis, Heather. 2015b. 'Plastic: Accumulation without Metabolism. In *Placing the Golden Spike: Landscapes of the Anthropocene*, edited by Dehlia Hannah and Sarah Krajewski, 66–73. University of Wisconsin-Milwaukee.

Davis, Heather and Etienne Turpin, eds. 2015. *Art in the Anthropocene: Encounters among Aesthetics, Politics, Environments and Epistemologies*. London: Open Humanities Press.

Decker, Julie. 2020. 'Climate of Change'. *Museum Management and Curatorship*, 35 (6), 636–652. DOI: 10.1080/09647775.2020.1836999

Deem, Alexandra. 2019. 'Mediated Intersections of Environmental and Decolonial Politics in the No Dakota Access Pipeline Movement'. *Theory, Culture & Society*, 36 (5), 113–131. DOI: 10.1177/0263276418807002

De La Cadena, Marisol. 2019. 'Uncommoning Nature: Stories from the Anthropo-Not-Seen'. In *Anthropos and the Material*, edited by Penny Harvey, Christian Krohn-Hansen and Knut G. Nustad, 35–58. Durham and London: Duke University Press.

De Landa, Manuel. (1997) 2019. *A Thousand Years of Nonlinear History*. New York: Zone Books.

Deleuze, Gilles. 1994. *Difference and Repetition*. Translated by Paul Patton. New York: Columbia University Press.

Deleuze, Gilles and Félix Guattari. (1980) 2004. *A Thousand Plateaus*. Translated by Brian Massumi London and New York: Continuum.

DeLoughrey, Elizabeth M. 2019. *Allegories of the Anthropocene*. Durham and London: Duke University Press.

Demos, T.J. 2016. *Decolonizing Nature: Contemporary Art and the Politics of Ecology*. Berlin: Sternberg Press.

Demos, T.J. 2017. *Against the Anthropocene: Visual Culture and Environment Today*. Berlin: Sternberg Press.

Descola, Philippe. 2013. *The Ecology of Others*. Translated by Geneviève Godbout and Benjamin P. Luley. Chicago: Prickly Paradigm Press.

DeSilvey, Caitlin. 2017. *Curated Decay: Heritage Beyond Saving*. Minneapolis and London: University of Minnesota Press.

Deutsches Museum website. www.deutsches-museum.de/en/exhibitions/humanity-and-environment/environment/. Accessed 23 April 2021.

Dibley, Ben. 2017. 'Prospects for a Common World Museums, Climate Change, Cosmopolitics'. In *Climate Change and Museum Futures*, edited by Fiona R. Cameron and Brett Neilson, 34–50. New York and London: Routledge.

Dredge Research Collaborative. 2013. 'Packaging Sludge and Silt'. In *Making the Geologic Now: Responses to Material Conditions of Contemporary Life*, edited by Elizabeth Ellsworth and Jamie Kruse, 72–78. New York: Punctum Books.

Drohan, Patrick J., John L. Havlin, J. Patrick Megonigal and H.H. Cheng. 2010. 'The "Dig It!" Smithsonian Soils Exhibition: Lessons Learned and Goals for the Future'. *SSAJ*, 74 (3) May–June, 697–705.

Drubay, Diane and Asha Singhal. 2020. 'Dialogue as a Framework for Systemic Change'. *Museum Management and Curatorship*, 35 (6), 663–670. DOI:10.1080/09647775.2020.1837001

Dudley, Sandra H. 2015. 'What, or Where, Is the (Museum) Object? Colonial Encounters in Displayed Worlds of Things'. In *Museum Theory*, edited by Andrea Witcomb and Kylie Message. *The International Handbook of Museums Studies*. General editors, Sharon Macdonald and Helen Rees Leahy, 41–62. Chichester, West Sussex: Wiley Blackwell.

Ellsworth, Elizabeth and Jamie Kruse, eds. 2013. *Making the Geologic Now: Responses to Material Conditions of Contemporary Life*. New York: Punctum Books.

Fisher, Mark. 2009. *Capitalist Realism: Is There No Alternative?* Winchester: O Books.

Flannery, Tim. 2020. *The Climate Cure: Solving the Climate Emergency in the Era of COVID-19*. Melbourne: Text Publishing.

Foucault, Michel. (1966) 1989. *The Order of Things: An Archaeology of the Human Sciences*. London and New York: Routledge.

Fox, William L. 2013. 'From Rock Art to Land Art/From Pleistocene to Anthropocene'. In *Making the Geologic Now: Responses to Material Conditions of Contemporary Life*, edited by Elizabeth Ellsworth and Jamie Kruse, 42–45. New York: Punctum Books.

Fox, William L. 2017. 'The Art of the Anthropocene'. In *Curating the Future: Museums, Communities and Climate Change*, edited by Jennifer Newell, Libby Robin and Kirsten Wehner, 196–205. Abingdon and New York: Routledge.

Fredengren, Christina and Cecilia Åsberg. 2020. 'Checking in With Deep Time : Intragenerational Care in Registers of Feminist Posthumanities, the case of Gärstadsverken'. In *Deterritorializing the Future: Heritage in, of and after the Anthropocene*, edited by Rodney Harrison and Colin Sterling, 56–95. London: Open Humanities Press.

Fritsch, Matthias, Philippe Lynes and David Wood. 2018. 'Introduction'. In *Eco-Deconstruction: Derrida and Environmental Philosophy*, 1–26. New York: Fordham University Press.

Gaitán Ammann, Felipe. 2006. 'Golden Alienation: The Uneasy Fortune of the Gold Museum in Bogotá. *Journal of Social Archaeology*, 6 (2), 227–254. DOI: 10.1177/1469605306064242

Gaskill, Malcolm. 2021. 'Philosophical Vinegar, Marvellous Salt'. *London Review of Books*, Vol. 43, 14, 15 July.

Gaynor, Andrea and Ian McLean. 2005. 'The Limits of Art History: Towards and Ecological History of Landscape Art'. *Landscape Review*, 11 (1), 4–14.

Ghosh, Amitav. 1992. 'Petrofiction'. *The New Republic*, March 2, 29–34.

Ghosh, Amitav. 2016. *The Great Derangement: Climate Change and the Unthinkable*. Chicago and London: The University of Chicago Press.

Gibson, Ross. 2006. 'Spirit House'. *South Pacific Museums: Experiments with Culture*, edited by Chris Healy and Andrea Witcomb, 23.1–23.6. Melbourne: Monash University Press.

Gronholt-Pederson, Jacob. 2021. 'Greenland Prepares Legislation to Halt Kuannersuit Rare Earth Mine', *Arctic Today*. September 11. www.arctictoday.com/greenland-prepares-legislation-to-halt-kuannersuit-rare-earth-mine/

Guattari, Félix. 1995. *Chaosmosis: An Ethico-aesthetic Paradigm*. Translated by Paul Bains and Julian Pefanis. Sydney: Power Publications.

Haigney, Sophie. 2021. 'The Challenge of Making an Archive of the Climate Crisis'. *The New Yorker*, October 4. www.newyorker.com/culture/annals-of-inquiry/the-challenge-of-making-an-archive-of-the-climate-crisis

Halperin, Ilana. 2015. 'Physical Geology/The Library'. In *Art in the Anthropocene: Encounters Among Aesthetics, Politics, Environments and Epistemologies*, edited by Heather Davis and Etienne Turpin, 79–84. London: Open Humanities Press.

Hamilton, Clive. 2007. *Scorcher: The Dirty Politics of Climate Change*. Melbourne: Black Inc.

Hamilton, Clive and Richard Denniss. 2005. *Affluenza: When Too Much Is Never Enough*. Sydney: Allen & Unwin.

Hannah, Dehlia. 2015. 'One Spike after Another: Site Specificity and Anthropocene Stratigraphy.' In *Placing the Golden Spike: Landscapes of the Anthropocene*, edited by Dehlia Hannah and Sarah Krajewski, 20–25. Milwaukee: University of Wisconsin-Milwaukee.

Hannah, Dehlia and Sarah Krajewski. 2015. *Placing the Golden Spike: Landscapes of the Anthropocene*. Milwaukee: University of Wisconsin-Milwaukee.

Hansen, Christine. 2017. 'Programming Interlude 1: Curating Fire'. In *Climate Change and Museum Futures*, edited by Fiona R. Cameron and Brett Neilson, 127–131. New York and London: Routledge.

Hansen, Christine. 2018. 'Deep Time and Disaster: Black Saturday and the Forgotten Past. *Environmental Humanities*, 10 (1/May), 226–240.

Haraway, Donna J. 2004. 'Teddy Bear Patriarchy: Taxidermy in the Garden of Eden, New York City, 1908-1936'. In *Grasping the World: The Idea of the Museum*, edited by Donald Preziosi and Claire Farago, 242–249. Aldershot and Burlington: Ashgate.

Haraway, Donna J. 2016. *Staying with the Trouble: Making Kin in the Chthulucene*. Durham and London: Duke University Press.

Harkinson, Josh. 2012. 'Science museums celebrate the wonder of … fracking:!' *Mother Jones*, December 3. www.motherjones.com/politics/2012/12/perot-hamm-science-museum-fracking/

Harkness, Rachel, Cristián Simonetti and Judith Winter. 2018. 'Concretes Speak: A Play in One Act'. In *Future Remains: A Cabinet of Curiosities for the Anthropocene*, edited by Gregg Mitmann, Marco Armiero and Robert S. Emmett, 29–39. Chicago and London: The University of Chicago Press.

Harrison, Rodney. 2015. 'Beyond "Natural" and "Cultural" Heritage: Toward an Ontological Politics of Heritage in the Age of Anthropocene'. *Heritage & Society*, 8 (1), 24–42. DOI: 10.1179/2159032X15Z.00000000036

Hassan, Robert and Thomas Sutherland. 2017. *Philosophy of Media*. London and New York: Routledge.

Hayman, Richard. 2016. *Coal Mining in Britain*. Oxford: Shire Publishing.

Hayman, Richard and Wendy Horton. 2009. *Ironbridge History and Guide*. Gloucestershire: The History Press.

Henning, Michelle. 2006. *Museums, Media and Cultural Theory*. Berkshire: Open University Press.

Hetherington, Kevin. 1997. 'Museum Topology and the Will to Connect'. *Journal of Material Culture*, 2 (2), 199–218.

Hird, Myra J. 2017. 'Proliferation, Extinction, and an Anthropocene Aesthetic'. In *Posthumous Life: Theorizing Beyond the Posthuman*, edited by Jami Weinstein and Claire Colebrook, 251–269. New York: Columbia University Press.

Howarth, Robert W, Renee Santoro and Anthony Ingraffea. 2011. 'Methane and the Greenhouse-Gas Footprint of Natural Gas from Shale Formations: A Letter'. *Climate Change*, 106, 679–690. DOI: 10.1007/s10584-011-0061-5

Hulme, Mike. 2009. *Why We Disagree about Climate Change: Understanding Controversy, Inaction and Opportunity*. Cambridge, UK: Cambridge University Press.

Hulme, Mike. 2017. 'Why We *Should* Disagree about Climate Change'. In *Climate Change and Museum Futures*, edited by Fiona R. Cameron and Brett Neilson, 9–15. New York and London: Routledge.

Huntley, Rebecca. 2020. *How to Talk about Climate Change in a Way That Makes a Difference*. Murdoch Books.

Ilyas, Sadia and Jae-chun Lee. 2018. *Gold Metallurgy and the Environment*. Milton: CRC Press.

Ingold, Tim. 2012. 'Toward an Ecology of Materials'. *Annual Review of Anthropology*, 41, 427–442.

Ingold, Tim. 2021. *Correspondences*. Cambridge, UK: Polity.

IPCC. 2021. Sixth Assessment Report (AR6). www.ipcc.ch/assessment-report/ar6/

Isager, Lotte, Line Vestergaard Knudsen and Ida Theilade. 2021. 'A New Keyword in the Museum: Exhibiting the Anthropocene'. *Museum & Society*, 19 (1), 88–107.

James Hutton Founder of Modern Geology. n.d. Lothian and Borders GeoConservation. Edinburgh Geological Society. https://edinburghgeolsoc.org/downloads/James-Hutton-LBGC-leaflet.pdf

James Hutton Memorial. 1947. *Nature*, 159, 737. https://doi.org/10.1038/159737b0

Janes, Robert R. 2015. 'The End of Neutrality: A Modest Manifesto'. *Informal Learning Review*, 135, November/December, 3–8.

Janes, Robert R. 2020. 'Museums in Perilous Times'. *Museum Management and Curatorship*, 35 (6), 587–598. https://doi.org/10.1080/09647775.2020.1836998

Janes, Robert R. and Richard Sandell. 2019. *Museum Activism*. London and New York: Routledge.

Jeffery, Tom. 2021. 'Critical Realist Philosophy and the Possibility of an Eco-decolonial Museology'. *Museum & Society*, 19 (1), 48–70.

Jemisin, N.K. 2015. *The Fifth Season*. London: Orbit.

Jevons, William Stanley. (1865) 2013. 'The Coal Question'. In *The Future of Nature: Documents of Global Change*, edited by Libby Robin, Sverker Sörlin and Paul Warde, 78–84. Yale University Press.

Johns, Elizabeth. 1998. 'Landscape Painting in America and Australia in an Urban Century'. In *New Worlds From Old: 19th Century Australian & American Landscapes*. Hartford: National Gallery of Australia, Canberra and Wadsworth Atheneum.

Jones, Tod. 2015. 'Separate but Unequal: The Sad Fate of Aboriginal Heritage in Western Australia'. *The Conversation*. December 7. https://theconversation.com/separate-but-unequal-the-sad-fate-of-aboriginal-heritage-in-western-australia-51561

Jordan, John. 2021. 'The Work of Art in the Age of Extinction: Notes toward an Art of Aliveness'. In *The Routledge Companion to Contemporary Art, Visual Culture, and Climate Change*, edited by T.J. Demos, Emily Eliza Scott and Subhankar Banerjee, 389–398. New York: Routledge.

Joseph, Richard. 2015. 'The Cost of Managerialism in the University: An Autoethnographical Account of an Academic Redundancy Process'. *Prometheus*, 33 (2), 139–163. DOI: 10.1080/08109028.2015.1092213

Karp, Ivan and Fred Williams. 1996. 'Constructing the Spectacle of Culture in Museums'. In *Thinking about Exhibitions*, edited by Reesa Greenberg, Bruce W. Ferguson and Sandy Nairne, 251–267. London and New York: Routledge.

Katie Paterson Artwork. http://katiepaterson.org/portfolio/campo-del-cielo/. Accessed 11 November 2010.

Kennedy Grimsted, Patricia, ed. 2015. *Archives in Russia: A Directory and Bibliographic Guide to Holdings in Moscow and St. Petersburg*. London and New York: Routledge.

Keogh, Luke and Nina Möllers. 2017. 'Pushing Boundaries: Curating the Anthropocene at the Deutsches Museum, Munich'. In *Climate Change and Museum Futures*, edited by Fiona R. Cameron and Brett Neilson, 78–89. New York and London: Routledge.

Kim, Jin-Woo, Zhong Lu and Kimberly Degrandpre. 2016. 'Ongoing Deformation of Sinkholes in Wink, Texas, Observed by Time-Series Sentinel-1A SAR Interferometry (Preliminary Results)'. *Remote Sensing*, 8 (4).

Kingsolver, Barbara. 2012. *Flight Behaviour*. London: Faber and Faber.

Klein, Naomi. 2015. *This Changes Everything: Capitalism vs. The Climate*. Penguin Books.

Kollar, Albert D. 2021. Blog. Understanding Fossil Fuels through Carnegie Museums' Exhibits. May 5. https://carnegiemnh.org/understanding-fossil-fuels-through-carnegie-museums-exhibits/

Larsen, Janike Kampevold. 2013. 'Imagining the Geologic'. In *Making the Geologic Now: Responses to Material Conditions of Contemporary Life*, edited by Elizabeth Ellsworth and Jamie Kruse, 83–89. New York: Punctum Books.

Larsen, Lars Bang. 2014. '1000 Words: Katie Paterson and Margaret Atwood', *Artforum International*, 53 (3), 262–263.

Lascelles, Bruce. 2015. 'What Is Soil?' In *Soil Culture: Bringing the Arts Down to Earth*, edited by Clive Adams and Daro Montag, 11–16. Devon: Centre for Contemporary Art and the Natural World.

Latour, Bruno. 1993. *We Have Never Been Modern*. Translated by Catherine Porter. Hertfordshire: Harvester Wheatsheaf.

Latour, Bruno. 2017. *Down to Earth: Politics in the New Climatic Regime*. Translated by Catherine Porter. Cambridge, UK: Polity.

Laurence, Alison. 2022. 'Out of Time at the La Brea Tar Pits: People and Other Animals in a Time Capsule of Ice Age Los Angeles'. *Museum & Society*, July, 20 (1), 71–88.

LeMenager, Stephanie. 2016. *Living Oil: Petroleum Culture in the American Century*. New York: Oxford University Press.

Lingis, Alphonso. 2018. 'Gold'. *Cultural Politics*, 14 (1), March, 55–62.

Lynch, Bernadette. 2010. 'Introduction – neither helpful nor unhelpful – a clear way forward for the useful museum'. In *Museums and Social Change: Challenging the Unhelpful Museum*, edited by Adele Chynoweth, Bernadette Lynch, Klaus Peterson and Sarah Smed. London: Routledge.

Lyons, Steve and Kai Bosworth. 2019. 'Museums in the Climate Emergency'. In *Museum Activism*, edited by Robert R. Janes and Richard Sandell, 174–185. London and New York: Routledge.

Macdonald, Graeme. 2017. "Monstrous Transformer": Petrofiction and World Literature. *Journal of Postcolonial Writing*, 53 (3), 289–302. DOI: 10.1080/17449855.2017.1337680

Macfarlane, Robert. 2019. *Underland: A Deep Time Journey*. Penguin Books.

References

Mahalia Dobson and Petra Cahill. 2021. 'Australia's dismal climate record comes under COP26 spotlight'. 25 November. www.cnbc.com/2021/11/26/australias-dismal-climate-record-comes-under-cop26-spotlight.html

Mahoney, Emma. 2021. 'From Institutional to Interstitial Critique: The Resistant Force that is Liberating the Neoliberal Museum from Below'. In *The Routledge Companion to Contemporary Art, Visual Culture, and Climate Change*, edited by T.J. Demos, Emily Eliza Scott and Subhankar Banerjee, 409–417. New York: Routledge.

Main, George. 2017. 'Food Stories for the Future'. In *Curating the Future: Museums, Communities and Climate Change*, edited by Jennifer Newell, Libby Robin and Kirsten Wehner, 171–180. London and New York: Routledge.

Malm, Andreas. 2016. *Fossil Capital: The Rise of Steam-Power and the Roots of Global Warming*. London: Verso.

Mann, Geoff. 2022. 'Reversing the Freight Train'. *London Review of Books*, 44 (16), 18 August.

Mann, Sally. 2003. *What Remains*. Boston: Bullfinch Press.

Marder, Michael. 2018. 'Ecology as Event'. In *Eco-Deconstruction: Derrida and Environmental Philosophy*, edited by Matthias Fritsch, Philippe Lynes and David Wood, 141–164. New York: Fordham University Press.

Maslin, Mark. 2014. *Climate Change: A Very Short Introduction*. Oxford: Oxford University Press.

McKay, Don. 2013. 'Ediacaran and Anthropocene: Poetry as a Reader of Deep Time'. In *Making the Geologic Now: Responses to Material Conditions of Contemporary Life*, edited by Elizabeth Ellsworth and Jamie Kruse, 46–54. New York: Punctum Books.

McKenzie, Bridget. 2020. 'Climate Museums UK: A Contemporary Response to the Earth Crisis'. *Museum Management and Curatorship*, 35 (6), 671–683. DOI: https://doi.org/10.1080/09647775.2020.1837003

Message, Kylie and Eleanor Foster. 2019. 'Museum Activism', *Museum Management and Curatorship*, 34 (6), 617–622. DOI: 10.1080/09647775.2019.1639346

Miller, Daegan. 2018. 'On Possibility or, The Monkey Wrench'. In *Future Remains: A Cabinet of Curiosities for the Anthropocene*, edited by Gregg Mitman, Marco Armiero and Robert S. Emmett, 143–148. Chicago and London: The University of Chicago Press.

Miller, Toby. 2015. 'Museums, Ecology, Citizenship'. In *Museum Theory*, edited by Andrea Witcomb and Kylie Message. *The International Handbook of Museums Studies*, 139–156. General editors, Sharon Macdonald and Helen Rees Leahy. Chichester, West Sussex: Wiley Blackwell.

Mirzoeff, Nicholas. 2015. *How to See the World*. UK: Penguin Books.

Mirzoeff, Nicholas. 2018. 'It's Not the Anthropocene, It's the White Supremacy Scene: or, The Geological Color Line'. In *After Extinction*, edited by Richard Grusin, 123–149. Minneapolis and London: University of Minnesota Press.

Mitman, Gregg, Marco Armiero and Robert S. Emmett, eds. 2018. *Future Remains: A Cabinet of Curiosities for the Anthropocene*. Chicago and London: The University of Chicago Press.

Montag, Daro. 2015. 'Speaking of Soil … For Soil thou Art'. In *Soil Culture: Bringing the Arts down to Earth*, edited by Clive Adams and Daro Montag, 19–32. Devon: Centre for Contemporary Art and the Natural World.

Morton, Timothy. 2013. *Hyperobjects: Philosophy and Ecology after the End of the World*. Minneapolis: University of Minnesota Press.

Muecke, Stephen. 2014. 'Global Warming and Other Hyperobjects'. *Los Angeles Review of Books*. February 20.

Murphy, Padraig, Pat Brereton and Fiachra O'Brolchain. 2021. 'New Materialism, Object-Oriented Ontology and Fictive Imaginaries: New Directions in Energy Research'. In *Energy Research & Social Science*, 79, September. https://doi.org/10.1016/j.erss.2021.102146

Murray, Mitch R. 2020. 'David Mitchell's Storytelling and the Metalife of Utopia', *ASAP/Journal*, 5 (1), 181–202.

Museums Association website. Museums for Climate Justice Campaign. (www.museumsassociation.org/campaigns/museums-for-climate-justice). Accessed 8 October 2022.

Nayar, Pramod. 2014. *Posthumanism*. Cambridge, UK: Polity Press.

Negarestani, Reza. 2008. *Cyclonopedia: Complicity with Anonymous Materials*. Melbourne: re-press.

Nietzsche, Friedrich. (1874) 1997. *Untimely Meditations*, edited by Daniel Breazeale, translated by R.J. Hollingdale. Cambridge, UK: Cambridge University Press.

Nixon, Rob. 2018. 'The Anthropocene: The Promise and Pitfalls of an Epochal Idea'. In *Future Remains: A Cabinet of Curiosities for the Anthropocene*, edited by Gregg Mitman, Marco Armiero and Robert S. Emmett, 1–18. Chicago and London: The University of Chicago Press.

Nummedal, Tara E. 2011. 'Words and Works in the History of Alchemy'. *Isis*, 102 (2), 330–337.

Nuttall, Mark. 2013. 'Zero Tolerance, Uranium and Greenland's Mining Future'. *Polar Journal*, 3 (2), 368–383.

Oreskes, Naomi and Erik M. Conway. 2014. *The Collapse of Western Civilization: A View from the Future*. New York: Columbia University Press.

Osborne, Erika. 2013. 'Exposing the Anthropocene: Art and Education in the "Extraction State"'. In *Making the Geologic Now: Responses to Material Conditions of Contemporary Life*, edited by Elizabeth Ellsworth and Jamie Kruse, 62–65. New York: Punctum Books.

Ovid. 2004. *Metamorphoses*. Translated by David Raeburn. London: Penguin.

Oxfam International. 2016. '62 people own the same as half the world', 18 January. www.oxfam.org/en/press-releases/62-people-own-same-half-world-reveals-oxfam-davos-report

Parikka, Jussi. 2015. *A Geology of Media*. Minneapolis and London: University of Minnesota Press.

Parliament of Australia. 2011. Inquiry into the use of 'fly-in, fly-out' (FIFO) workforce practices in regional Australia. www.aph.gov.au/parliamentary_business/committees/house_of_representatives_committees?url=ra/fifodido/report.htm

Pascoe, Bruce. 2014. *Dark Emu: Aboriginal Australia and the Birth of Agriculture*. Broome: Magabala Books.

Patrizio, Andrew. 2019. *The Ecological Eye: Assembling an Ecocritical Art History*. Manchester: Manchester University Press.

Pearce, Susan M. 2010. 'Foreword'. In *Museum Materialities: Objects, Engagements, Interpretations*, edited by Sandra H. Dudley, xiv–xix. London and New York: Routledge.

Pearse, Guy. 2009. 'Quarry Vision: Coal, Climate Change and the End of the Resources Boom'. *Quarterly Essay*, (33), 1–122.

Pétursdóttir, Þóra and Bjørnar Olsen. 2014. 'An Archaeology of Ruins'. In *Ruin Memories: Materialities, Aesthetics and the Archaeology of the Recent Past*, edited by Bjørnar Olsen and Þóra Pétursdóttir, 3–29. London: Routledge.
Phillips, Leigh. 2015. *Austerity Ecology & the Collapse-Porn Addicts*. Winchester: Zero Books.
Povinelli, Elizabeth A. 2016. *Geontologies: A Requiem to Late Liberalism*. Durham and London: Duke University Press.
Richards, Jim. 2016. *Gold Rush: How I Made, Lost and Made a Fortune*. Fremantle: Fremantle Press.
Richer-de-Forges, Anne C., David J. Lowe and Budiman Minasny. 2021. 'A Review of the World's Soil Museums and Exhibitions'. *Advances in Agronomy*, 166, 277–304.
Robin, Libby, Dag Avengo, Luke Keogh, Nina Möllers, Bernd Scherer and Helmuth Trischler. 2014. 'Three Galleries of the Anthropocene'. *The Anthropocene Review*, 1 (3), 207–224. DOI: 10.1177/2053019614550533
Robin, Libby, Dag Avengo, Luke Keogh, Nina Möllers and Helmuth Trischler. 2017. 'Displaying the Anthropocene In and Beyond Museums'. In *Curating the Future: Museums, Communities and Climate Change*, edited by Jennifer Newell, Libby Robin and Kirsten Wehner, 252–266. London and New York: Routledge.
Robins, Claire. 2013. *Curious Lessons in the Museum: The Pedagogic Potential of Artists' Interventions*. Farnham: Ashgate.
Robinson, Jake M. and Ross Cameron. 2020. 'The Holobiont Blindspot: Relating Host-Microbiome Interactions to Cognitive Bias and the Concept of "Unwelt"'. *Frontiers in Psychology. Theoretical and Philosophical Psychology*, 11: 591071.
Robinson, Kim Stanley. 2012. *2312*. London: Orbit.
Robinson, Kim Stanley. 2020. *The Ministry for the Future*. London: Orbit.
Rossée, Carlina, Wanuri Kahiu and Peter K. Haff. 2019. 'Future Worlds: Intelligent Soil, Technospheric Colonization, and a Habitat of Emotional Particles'. In *Field to Palette: Dialogues on Soil and Art in the Anthropocene*, edited by Toland, Alexander, Jay Stratton Noller and Gerd Wessolek, 429–442. Boca Raton: CRC.
Royal Society. (1788). Philosophical Transactions. 31 December. 'An account of a mass of native iron, found in South-America. By Don Michael Rubin de Celis. Communicated by Sir Joseph Banks, Bart. P.R.S'. https://royalsocietypublishing.org/doi/10.1098/rstl.1788.0004
Russell, Andrew and Lee Vinsel. 2006. 'Hail the Maintainers'. *Aeon*. 5 May.
Rutherford-Morrison, Lara. 2015. 'Playing Victorian: Heritage, Authenticity, and Make-Believe in Blists Hill Victorian Town, the Ironbridge Gorge'. *The Public Historian*, 37 (3), 76–101.
Salazar, Juan F. 2017. 'Futuring Global Change in Science Museums and Centers: A Role for Anticipatory Practices and Imaginative Acts'. In *Climate Change and Museum Futures*, edited by Fiona R. Cameron and Brett Neilson, 90–108. New York and London: Routledge.
Saldanha, Arun. 2017. 'Geophilosophy, Geocommunism: Is There Life after Man?' In *Posthumous Life: Theorizing Beyond the Posthuman*, edited by Jami Weinstein and Claire Colebrook, 225–247. New York: Columbia University Press.
Saraceno, Tomás, Sasha Engelmann and Bronislaw Szerszynski. 2015. 'Becoming Aerosolar: From Solar Sculptures to Cloud Cities'. In *Art in the Anthropocene: Encounters Among Aesthetics, Politics, Environments and Epistemologies*, edited by Heather Davis and Etienne Turpin, 57–62. London: Open Humanities Press.

Science Alert. 2014. 'Game of Thrones Exposed; How molten gold would really kill you'. 16 June. www.sciencealert.com/game-of-thrones-exposed-how-would-molten-gold-really-kill-you. Accessed 29 November 2019.

Seeger, Philip A., William A. Fowler and Donald D. Clayton. 1964. 'Nucleosynthesis of Heavy Elements by Neutron Capture'. *The Astrophysical Journal Supplement Series*, 11, 121.

Shaviro, Steven. 2014. 'Non-Phenomenological Thought'. In *Speculations V: Aesthetics in the Twenty-First Century*, edited by Ridvan Askin, Paul J. Ennis, Andreas Hagler and Philipp Schweighauser, 40–56. New York: Punctum Books.

Sheldrake, Merlin. 2020. *Entangled Life: How Fungi Make Our Worlds, Change Our Minds, and Shape Our Futures*. London: The Bodley Head.

Shelton, Anthony Alan. 2013. 'Museums and Museum Displays'. In *Handbook of Material Culture*, edited by Christopher Tilley, Webb Keane, Susanne Küchler, Michael Rowlands and Patricia Spyer, 480–499. Los Angeles: Sage.

Simon, Nina. 2010. *The Participatory Museum*. Santa Cruz: Museum 2.0.

Skinner, Anne. 2014. *KCGM Kalgoorlie Consolidated Gold Mines: A Celebration of 25 Years*. Boulder: KCGM.

Smith, Hannah. 2021. 'The Story of Oil in Western Pennsylvania: What, How, and Why?' *Blog Carnegie Museum of Natural History*, May 17. https://carnegiemnh.org/the-story-of-oil-in-western-pennsylvania/

Smith, Laurajane. 2006. *Uses of Heritage*. London and New York: Routledge.

Sontag, Susan. 1992. *The Volcano Lover*. London: Vintage.

Sovereign Hill Museums Association. 2013. *Sovereign Hill Ballarat Australia Official Souvenir Booklet*. Ballarat: The Sovereign Hill Museums Association.

Spaid, Sue. 2019. 'Artisanal Soil'. In *Field to Palette: Dialogues on Soil and Art in the Anthropocene*, edited by Alexandra R. Toland, Jay Stratton Noller and Gerd Wessolek, 35–50. Boca Raton, FL: CRC Press.

Sparrow, Jeff. 2021. *Crimes against Nature: Capitalism and Global Heating*. Melbourne and London: Scribe.

Stanford Encyclopedia of Philosophy. 2013. 'Francis Bacon'. Metaphysics Research Lab, CSLI, Stanford University.

Steffen, Will. 2013. 'Commentary. Paul J. Crutzen and Eugene F. Stoermer, "The Anthropocene"' (2000). *The Future of Nature: Documents of Global Change*, edited by Libby Robin, Sverker Sorlin and Paul Warde, 486–490. New Haven and London: Yale University Press.

Stephens, Beth, Annie Sprinkle and Frederick L. Kirschenmann. 2019 'Soil Lovers Unite for a Down and Dirty Q&A'. In *Field to Palette: Dialogues on Soil and Art in the Anthropocene*, edited by Alexandra R Toland, Jay Stratton Noller and Gerd Wessolek, 545–554. Boca Raton: CRC Press.

Sterling, Colin. 2020. 'Heritage as Critical Anthropocene Method'. In *Deterritorializing the Future: Heritage in, of and after the Anthropocene*, edited by Rodney Harrison and Colin Sterling, 188–218. London: Open Humanities Press.

Sutcliffe-Braithwaite, Florence. 2021. 'Tesco and a Motorway'. *London Review of Books*, 43, 17. 9 September.

Thacker, Eugene. 2010. *After Life*. Chicago and London: The University of Chicago Press.

The Natural History Museum. http://thenaturalhistorymuseum.org/about/. Accessed 8 July 2021.

The Ryder Movies for Moderns. 'Talking about Ecosex – in Art, Theory, Practice and Activism'. www.theryder.com/magazine/feature-articles/talkin-about-ecosex-in-art-theory-practice-and-activism/

Þórsson, Bergsveinn. 2020. 'Walking through the Anthropocene: Encountering Materialisations of the Geological Epoch in an Exhibition Space'. *Nordic Museology*, 1, S. 103–119.

Toadvine, Ted. 2018. 'Thinking After the World: Deconstruction and Last Things'. In *Eco-Deconstruction: Derrida and Environmental Philosophy*, edited by Matthias Fritsch, Philippe Lynes and David Wood, 50–80. New York: Fordham University Press.

Toland, Alexandra R, Jay Stratton Noller and Gerd Wessolek, eds. 2019. *Field to Palette: Dialogues on Soil and Art in the Anthropocene*. Boca Raton: CRC Press.

Tolvanen, Anne, Pasi Eilu, Artti Juutinen, Katja Kangas, Mari Kivinen, Mira Markovaara-Koivisto, Arto Naskali, Veera Solkannel, Sieji Tuulentie and Jukka Simia. 2019. 'Mining in the Arctic Environment – A Review from Ecological, Socioeconomic and Legal Perspectives'. *Journal of Environmental Management*, 233 (2019), 932–844.

Tøndborg, Britta. 2013. 'The Dangerous Museum: Participatory Practices and Controversy in Museums Today'. *Nordisk Museologi*, 2, 3–16.

Trigger, Rebecca and Samantha Goerling. 2021. 'WA Aboriginal Heritage Law Passes but Concerns Remain It Won't Prevent Another Juukan Gorge'. *ABC News*, 15 December. www.abc.net.au/news/2021-12-15/aboriginal-heritage-law-aimed-at-stopping-another-juukan-passes/100701120

Tsing, Anna, Heather Swanson, Elaine Gan and Nils Bubandt, eds. 2017. *Arts of Living on a Damaged Planet*. Minneapolis and London: University of Minnesota Press.

Tuan, Yi-Fu. 2014. *Space and Place: The Perspective of Experience*. Minneapolis and London: University of Minnesota Press.

Turpin, Etienne and Valeria Federighi, eds. 2013. 'A New Element, A New Force: Antonio Stoppani's Anthropozoic'. In *Making the Geologic Now: Responses to Material Conditions of Contemporary Life*, edited by Elizabeth Ellsworth and Jamie Kruse, 34–41. New York: Punctum Books.

UN Climate Change Conference UK. 2021. https://ukcop26.org/global-coal-to-clean-power-transition-statement/

Van Alphen, Ernst. 2001. 'Toys and Affect: Identifying with the Perpetrator in Contemporary Holocaust Art'. *Australian and New Zealand Journal of Art*, 2 (2–1), 158–189.

van Dooren, Thom. 2017. 'The Last Snail Loss, Hope and Care for the Future'. In *Curating the Future: Museums, Communities and Climate Change*, edited by Jennifer Newell, Libby Robin and Kirsten Wehner, 145–152. London and New York: Routledge.

Vernadsky, Paul. 2021. 'Don't Make a Fetish of Andreas Malm.' *Workers' Liberty* 4 May. www.workersliberty.org/story/2021-05-04/dont-make-fetish-andreas-malm

Viano, Lucas. 2015. 'Meteorite Thefts Pose a Problem in Ancient Impact Field'. *Scientific American*, June 19. www.scientific American.com/article/meteorite-thefts-pose-a-problem-in-ancient-impact-field/. Accessed 2 January 2020.

Weinstein, Jami and Claire Colebrook, eds. 2017. *Posthumous Life: Theorizing Beyond the Posthuman*. New York: Columbia University Press.

Weintraub, Karen. 2013. 'The New Gold Standard', *Nature, Outlook: Gold*, 495, 7440 (March 14).
Wickham, Dorothy. 2014. *Eureka*. Bakery Hill: BGS Publishing.
Wilkinson, Marian. 2020. *The Carbon Club*. Sydney: Allen and Unwin.
Willbold, Matthias, Tim Elliot and Stephen Moorbath. 2011. 'The Tungsten Isotopic Composition of the Earth's Mantle before the Terminal Bombardment'. *Nature*, 477 (7363), 195–198.
Williams, Raymond. 1976. *Keywords: A Vocabulary of Culture and Society*. London: Fontana Press.
Williams, Rosalind. 2008. *Notes on the Underground*. Cambridge: The MIT Press.
Witcomb, Andrea. 2010. 'Remembering the Dead by Affecting the Living: The case of a Miniature Model of Treblinka'. In *Museum Materialities: Objects, Engagements, Interpretations*, edited by Sandra H. Dudley, 39–52. London and New York: Routledge.
Wolfe, Cary. 2010. *What is Posthumanism?* Minneapolis: University of Minnesota Press.
Wray, Lynn. 2019. 'Taking a Position: Challenging the Anti-authorial Turn in Art Curating'. In *Museum Activism*, edited by Robert R. Janes and Richard Sandell, 315–325. London and New York: Routledge.
Yunkaporta, Tyson, 2021 'All Our Landscapes Are Broken: Right Story and the Law of the Land'. *Griffith Review*, 78. www.griffithreview.com/articles/all-our-landscapes-are-broken/
Yusoff, Kathryn. 2013. 'Geologic Life: Prehistory, Climate, Futures in the Anthropocene'. *Environment and Planning D: Society and Space*, 31, 779–795.
Yusoff, Kathryn. 2018. *A Billion Black Anthropocenes or None*. Minneapolis: University of Minnesota Press.
Yusoff, Kathryn. 2019. 'Geologic Realism: On the Beach of Geologic Time'. *Social Text*, 37 (1) (138), 1–26.
Zalasiewicz, Jan. 2010. *The Planet in a Pebble*. Oxford: Oxford University Press.
Žižek, Slavoj. 2017. 'Lessons from the "Airpocalypse"'. *In These Times*. January 10. https://inthesetimes.com/article/spaceship-earth-lessons-of-airpocalypse-slavoj-zizek-climate-ecology-smog

Index

abjection 24, 33, 67, 136–7, 142
activism 13, 59; and artists 46; and museums 16, 45, 51, 103
Adani, Gautan 97
Afrofuturism 26
Agricola, Georgius: *On the Nature of Metals (De Re Metallica)* 79, 131
agriculture 21, 22, 25, 33, 69; automation of 33; hyperaccumulator plants 25; stump-jump plough 21; *see also* soils
Alaska 58
alchemy 79–80
Allard, LaDonna Brave Bull 63
Almarcegui, Lara *Aushub aus Basel* 27
Amazon Forest deforestation 50, 135
American Museum of Natural History (AMNH) 16, 97, 98
Anchorage Museum 58
Antarctica 39, 45
Anthropocene 3, 15, 67, 140; aesthetic 30; Anthropocene Living Room 71; and art 12; and concrete 116; as keyword 68; beginning of 33, 47, 69, 142; discourse of 3, 68, 69; exhibitions 16, 35–9, 70, 73; objections to 33, 67–71; Slam 38–9; thesis 18, 35, 65, 70, 75; *see also Welcome to the Anthropocene* exhibition
Anthropocene Extinction exhibition 36
anthropocentricism 9, 27, 28, 61, 71–3, 82, 86, 91; death of 40; in museum displays 6
anti-humanism 57, 75, 138, 142; and art history 12; and museums 42
Arctic climate impacts 58–9; mining 104–5
Aristotle 80; hylomorphism 74–5
art 2, 4, 16–17, 36, 39, 83–4, 105, 121–2; Aboriginal 112, 115; *Art in the Anthropocene* (book) 12, 13; and fieldwork 98–9; Holocaust art 98; and landscape tradition 113–14, 116; role of 139; and soils 22–7; and the sublime 112; *see also* activism; Anthropocene; art history; individual artists
art history 12, 13, 35, 96, 112
Art Not Oil 46
Arts of Living on a Damaged Planet 12
Atmosphere: Exploring Climate Science exhibition 37, 42
Atmospherics exhibition 43
Attenborough, David 31
Australian Museum, the 42
Awdry, Wilbert Reverend 91

Bacon, Francis 79; *The New Atlantis* 80
bacteria 30, 31, 66, 108
Balkin, Amy 39, 45, 137
Barad, Karen 56–7, 75, 76
Beamish Museum 90, 93
Beck, Ulrich 48
Belisle, Brooke 83–4
Bennett, Jane 18, 29
Bennett, Tony: evolutionary museum, the 83
BHP Billiton 50, 97
Big Mining 7, 17, 24, 51, 90, 94, 102, 133
Big Tech 49, 51, 138
Bird Rose, Deborah 10
Bishop, Claire 53
Black Mesa, the 17
Blists Hill Victorian Town 84
Bootu Creek mine 11
Boozer, Margaret, *Correlation Drawing/Drawing Correlations* 22
Bostrom, Nick 40, 41
Boulton, Elizabeth 1–2, 5, 67

Index

Bourriaud, Nicholas 53
BP (British Petroleum) 17, 47, 103
Braidotti, Rosi 15; and nomad thought 75
British Museum, the 81
Buffon Declaration 52
burial practices 23, 24
Burnet, Thomas, *Sacred History of the Earth* 80, 81

California State Mining and Mineral Museum 126, 128
Cameron, Fiona 5, 7, 8, 58, 138; and Brett Neilson 3; critique of climate exhibitions 37, 42–4
carbon dioxide 20, 30, 33, 43, 64, 89, 119
car culture 88, 100, 102, 111
Capitaloscene 11, 46, 76, 102, 138
Carnegie Museum of Natural History 70–1, 95, 105
Center for Land Use Interpretation (CLUI) *Perpetual Architecture* 27–9
Chakrabarty, Dipesh 47
Chandler, Jo 95
Chin, Mel *Revival Field* projects 25
Chthuluscene *see* Haraway, Donna
Clark, Nigel 71
Clark, Timothy 31
climate change: activism 40, 44, 45; apathy 2, 13; archiving 45, 122, 137; blame 42; and capital realism 136; denial and inaction 2, 14, 52, 54, 91; displays 38, 42; emergency 1, 5, 51, 54, 64, 100; governance 1, 2, 3, 37, 64, 68, 89, 135; guilt 42–3, 111; guilt as control 65; health decline and death 5, 14, 135–6; irony 27; misinformation 14; science 1, 14, 22, 77, 89, 135; tipping points 135; *see also* soil
Climate Change: Our Future, Our Choice exhibition 42
Climate Change Wall exhibition 42
climate fiction literature (cli-fi) 3, 18, 64–5
Climate Museum U.K. (CMUK) 44, 46
coal 75, 89, 95; as actant 44; and Australia 96–7, 119–20, 133; exports 91, 97, 119; collieries 92, 93; COP26 Transition Statement on Global Coal to Clean Power 96; mining 17, 91, 94; and steam power 88
Colebrook, Claire 72–3, 140; and Jami Weinstein 41
Coalbrookdale by Night 84
Coalbrookdale Museum of Iron 84
Cohen, Jeffrey, 82, 86, 119; *Stone* (book) 8
collections 1, 6, 8, 39, 75, 93; anomalies 39, 60; artefacts 142; cabinet of curiosities 7, 39; challenging the ICOM legacy 44; deaccessioning libraries 49; gold 130; repurposing 78, 86, 136, 139; rock collections 60, 71, 80–1; of science 37–8, 83, 93; soil 20, 22; *see also* lapidaries
Columbia Central Bank 128
concrete 22, 27, 63, 116
Concretes Speak 116–17
connectivity discourse 7, 51, 69, 72; connectivism 52; in displays 36
consumer culture 2, 30, 86, 88, 91, 111, 138; and 'affluenza' 133; and neoliberalism 7
COP conferences 77, 96
Corriel, Eric, *Water Will be Here* 47
cosmopolitics 15
Countryside, the Future exhibition 58
COVID-19 pandemic 2, 80, 119
Creation, The (biblical) 80, 81, 86
critical archeology 27, 28, 62
critical heritage 5, 27, 62, 85
critical thinking 8, 14, 35, 40, 47 63, 71; demise in university teaching 49, 103
Crutzen, Paul J. 142; and Eugene Stoermer 67, 69
Cuno, James 52
Culp, Andrew 52, 72

Dakota Access Pipeline 10, 15
Davis, Heather 111; and Etienne Turpin, *Art in the Anthropocene* 12, 13, 139
death 63; and decomposition 23, 24; of the world 40, 72, 142
Death of the PostHuman Vol 1 72
Decker, Julie 58
Declaration on the Importance and Value of Universal Museums 52
deconstruction philosophy 61
Deem, Alexandra 63
Deep Roots exhibition 25
Deepwater Horizon disaster 47, 103
defamiliarization 13, 55, 98, 114
de la Cadena, Marisol 70
De Landa, Manuel 61

Deleuze, Gilles 66, 72, 76; and Félix Guattari: metallurgy 74; nomadology 74–5
DeLoughrey, Elizabeth 69
Demos T.J. 13, 14, 26, 70, 71; on dOCUMENTA (13) 26
Descola, Philippe 40
DeSilvey, Caitlin 62
Deutsches Museum 37, 94
de vries, hermann 22–3
Dibley, Ben 52
digital spectacle 49, 118, 119
Dig It! The Secrets of Soil exhibition 25
Dithering, The 64, 133
Dredge Research Collaborative, The 30–1
Drubay, Diane and Asha Singhal 35
dust 24, 32
Dynamic Earth exhibition 42, 119

Ecologic exhibition 42
eco-sexual weddings 23
Eldfell Volcano 71
Ellsworth, Elizabeth and Jamie Kruse 57, 68
Entangled Life (book) 34
Eureka Stockade 127; Eureka Centre, the 127
exhibitions: on the Anthropocene 4, 35; climate-related 3, 35, 54; exhibition guilt 65; geological 71; 'Hot Topic' exhibits 53; *see also* individual exhibitions
extinction 5, 26, 31, 64–6, 95, 123, 134; extinction of the 'self' 72–3; and museums 139
Exxon Mobil 17, 46, 110

Field Museum of Natural History, the 16, 47
Field to Palette (anthology) 25, 27
FIFO (fly-in/fly-out) workforce 108
Fimiston Open Pit (Super Pit) 129
fire 5, 6, 21, 72, 136, 137; in Australia 2, 135, 141; in California 135–6
First Nations knowledge 10–11, 15, 21, 63, 72, 83; and hope 141; Tjipel 115
First Nation peoples 63, 66, 78, 112, 132, 134: agriculture 21; appropriation of motifs 7; Belup Peoples and Bootu Creek mine destruction 11; Black Mesa, the 17; Belyuen 115; Dakota Access Pipeline 10, 16, 63;

destruction of country 9–11, 16, 18, 100; and cultural dislocation 125, 126, 132; Greenlanders 103, 105; Inuit 38, 104; Inuktitut, Baffin Island 59; Murujuga (Burrup Peninsula) 11, 17, 115; Navaho cosmology 17; Northern Cheyenne 17; Puutu Kunti Kurrama and Pinikura Peoples (PKKP) 9, 17; Seneca Nation 105; Standing Rock Sioux 10, 63; Wadjuk Noongar ix; Wadawurrung and Dja Dja Wurrung Peoples 127; *see also* Juukan Gorge shelters
Flannery, Tim 133, 135
floods 5, 21
flow 10, 74, 79
Fort Worth Museum of Science and History 110
fossil capital 3, 14, 68, 72, 88, 90, 91, 102, 103
Foucault, Michel 40, 63
Fox, William 13, 24
fracking 109, 111
Friedrich, Caspar David *Wanderer Above the Sea of Fog* 112
fungus 5, 20, 23, 24, 66; *Entangled Life* (book) 34; mycorrhizal networks 21, 33–4, 37
Future Remains (book) 12

Gaitán Ammann, Felipe 127–8, 130; 'golden alienation' 130
Game of Thrones 131
gardens 21, 25, 26
geo-inclusion *see* inclusion
geological 19, 69, 112; actants 3; life 61; time 27, 71, 82, 114; 'turn' 57
geontology 11, 63, 73, 138; Tjipel 115
geo-philia 4, 9
geopoetry 73, 113, 117, 131
George C. Page Museum of La Brea Discoveries 107; 'tar pits' 108
geo-tubing 30–1
Ghosh, Amitav 64, 102; on petrofiction 100
Giant 100
globalisation 15, 75, 76
gold 19, 74; environmental impact 132; fever 128–9; goldrushes 126–7; and global finance 127; pre-Hispanic 124, 130; as puzzling lure 19, 124; and toxic pollutants 125, 131–2; the 'vault' 128; Welsh gold 127–8

golden spike, the 69
Gold Museum (Museo del Oro), Bogotá 124, 130
Gold of Africa: Jewellery and Ornaments from Ghana, Cote d'Ivoire, Mali and Senegal in the Collection of the Barbier-Mueller Museum exhibition 124
Great Derangement, The 65
Greek myth 86
greenhouse gases 14, 21, 89, 119, 110
Greenland 45, 59, 103–4, 135; ice sheet 135
Greenpeace 97
greenwashing 46, 134, 138
Guattari, Félix 68

Haacke, Hans *MetroMobiltan* 46
Halperin, Ilana 13, 71
Hamilton, William Sir 81
Hansen, Christine 136–7
Haraway, Donna 15, 23, 24, 26, 97–8; Chthulucene 33, 68, 138; odd kin 75, 77; 'staying with the trouble' 46, 63
Harman, Graham 59
Harrison, Rodney 5
Hassan, Robert and Thomas Sutherland 84
heritage 28, 84, 89, 95, 100, 108; and affect 116; authorised heritage discourse 85; and coal mining 93; difficult 4; 'new' 5; *see also* critical heritage
Hetherington, Kevin 39
Hird, Myra 30
history 3, 27, 36, 88, 91, 93, 98; antiquarian 85, 89; critical 95, 99–101; and geological time 47; monumental 90
Houston Museum of Natural Science 16, 17
Hulme, Mike 14, 140
humanism 15, 28, 81, 86, 114; hyper-humanism 69–70; *see also* anti-humanism
humanities 69, 71; university courses closure 49, 103
Huntley, Rebecca 42, 45
Hurricane Sandy 45
Hutton, James 81–2
hylomorphism 74
hyperobject 5–6, 14, 59–60, 86

Iceland 71
Inca Empire 131

inclusion 3, 6; and the cinematic museum 56; and cultural identity 51; geo-inclusion 4, 18, 61, 73, 75, 141; neoliberal appropriation of 52, 54; and the new museology 48; participation as inclusion 52–3; and popularism 58
inclusive museum, the 4, 17, 37, 51–5; dilemma 4, 7, 12, 18; inclusion movement, the 53
Indonesian Gold: Treasures from the National Museum, Jakarta 124
industrial heritage 84, 92, 93
Industrial Revolution, the 3, 84, 88, 131, 134
Ingold, Tim 7, 48, 55, 74
inhuman, the 2, 4, 27, 82, 87; entanglements 31, 51; geography 78; life of 32; way of seeing 23
innovation 30, 98, 139; miner's safety lamp 92; rhetoric 7, 48, 101; as climate fix 3, 43; hype 48; versus maintenance 48–9, 135
Intergovernmental Panel on Climate Change (IPCC) 1, 77
Inuit 38
Inuktitut 59
iron 84, 85; Ironbridge 84, 93; Iron Bridge 84
irony 27, 28, 29
Isager, Lotte, exhibition research 36, 47, 70
Isidore of Seville 87

James Watt steam engine 88
Janes, Robert 8, 40, 54; and Richard Sandell 8, 46, 51
Jeffery, Tom 5, 8, 44
Jemisin, N. K. *Broken Earth* trilogy 66–7
Jevons, William Stanley 90
Johns, Elizabeth 113
Just Stop Oil 45
Juukan Gorge Shelters, destruction of 9, 16, 17

Kahiu, Wanuri, *Pumzi*, 25
Kant, Immanuel 57
Keep America Beautiful campaign 43
Kettels, Robert 115
keywords 15, 68, 75
Kingsolver, Barbara: *Flight Behaviour* (book) 65
Klein, Naomi 17
Koch, David 97

Lal, Rattan 20
Landa, Edward 32
landscape 10, 23, 112–14; engineered 31
lapidaries 86, 11; Peterborough 120
Larsen, Janike Kampevold 112–13
Latour, Bruno 14, 76, 139
leachate 22, 30, 68, 75
Lee, Jae Rhim, decompiculture 23; *Infinity Burial Project* 23–4
LeMenager, Stephanie 100, 107–9; *Living Oil* 100
Let's Talk About the Weather exhibition 47
Libera, Zbigniew *Lego Construction Camp Set* 98
libraries *see* collections
Liberate Tate 103
liquid museum 5, 37, 138
Lyell, Charles: *Principals of Geology* 83
Lynch, Bernadette 54, 55, 58

Macdonald, Graeme 100, 102
Macfarlane, Robert 59; *Underland* (book) 33–4
Mahoney, Emma 103
Main, George 20–1
Making the Geologic Now (book) 29, 30, 68
Malm, Andreas 14, 89, 91, 96; *Fossil Capital: The Rise of Steam-Power and the Roots of Global Warming* (book) 88
Malmo Museum, The 53
managerialism 7, 48
Mann, Sally, *What Remains* 23
Mars 11, 14, 20
Maryland Historical Society 16
Massie, Miranda 45
material thinking 2, 3, 11, 116
McKay, Don 113
meteors 24, 118; Campo Del Cielo crater field 120 (*see* Paterson, Katie); Murchison meteorite 120
Metropolitan Museum of Art 46
Miller, Daegan 39
Mining the HMNS 16
Mining the Museum 16
Mirzoeff, Nicholas 32; and White Supremacy Scene 67
monkey wrench 38–9; as Earth First logo 39; *The Monkey Wrench Gang* (book) 39
Morton, Timothy 5, 59, 67; hyperobject 5–6, 60, 85
Mouffe, Chantal 53
Mullion Cove, Cornwell 62

Museu do Amajha/Museum of Tomorrow 49–50
Museum Activism (book) 51
museums: authority of 1, 5, 16, 31, 141; and climate change governance 2, 3, 8, 37, 51, 89; and contradiction 117; digital and online 44; and hubris 12, 42; and humanism 9; and innovation 49; inverted 112; myth of neutrality 51, 54; and neoliberalism 7, 65; and neutral stance 1, 3, 6, 8, 9, 36, 100; objects and memory 116; and participation 53; sponsorship 46, 55, 91, 97–8, 109, 110; theory 4, 5; and transformation 1, 19, 35, 48–9; *see also* collections; inclusive museum; liquid museum; objects
Museums Association (MA) UK, Museums for Climate Justice Campaign 141
Museum of Capitalism 4
Museum of Nonhumanity 4
Museum Management and Curatorship (journal) 44
Murujuga (Burrup Peninsula) rock sites 9, 11, 115

National History Museum, London 7; merchandising 7; Strategy to 2031 7
National Museum of Australia (NMA) 21
National Museum of Natural History (Smithsonian) 25
National Museum of Scotland 82
Natural History Museum, (NHM) Not an Alternative 16, 17
natureculture 23, 35, 44, 61
Negarestani, Reza: *Cyclonopedia* (book) 106
neo-liberalism 5, 18, 44, 49, 51, 76, 100, 138; and economics 14; museum and practices 5, 8, 65, 135; and universities 7, 75
new materialism 17, 25, 62, 74
new museology, the 7, 48, 54
New York Climate Museum 45
Nietzsche, Friedrich 90, 101
9/11 68, 70
Nixon, Rob 68, 70
No Dakota Access Pipeline (#NoDAPL) 63
nuclear fallout 68; CLUI 28–9; waste 29–30, 59; Yucca Mountain, Nevada 29

162 Index

object-oriented ontology 17, 60
objects 5, 28; in cinema 56; lure of 60, 72, 82; object-oriented ontology 17, 60; *see also* collections; museums
Ocean Star Offshore Drilling Rig and Museum 101, 109
odd kin 46, 75, 76, 138
oil 5, 74, 100; and petroleum 88, 92
Oil Sands Discovery Center and Heritage Park 108
Oreskes, Naomi and Erik Conway 2–3
Orford Ness lighthouse 62
organic chauvinism 9, 61
Ovid 86

Paddock Report, The 21
Paglin, Trevor *Artifacts* 83–4
palliative curation 27, 62
Paris Agreement 64, 94
participation 4, 38, 53–4
Participatory Museum, The 53
Paterson, Katie: *Campo del Cielo*/*Field of the Sky* 121–2; *Future Library* 122–3
Patrizio, Andrew 12, 96
Pearce, Susan 116
pedology 20, 22
People's Archive of Sinking and Melting, A 39, 45
permafrost 5, 20; melting 58
Perot Museum of Nature and Science 110
petrofiction 100, 102
petro-invisibility 100–1
petrotopia 100, 105
Pétursdóttir, Þóra and Bjørnar Olsen 27, 28
Phillips, Leigh 76
Pilbara, the 9, 11, 16, 115
Placing the Golden Spike: Landscapes of the Anthropocene exhibition 16–17, 47
Planet in a Pebble, The 31–2
plastics 13, 22, 75, 111
plastiglomerates 22, 68
Plato 73–4
Plumwood, Val 21
populism 49, 53, 58; and spectacle 107
PostHuman City, The exhibition 23–4
posthuman (isms) 18, 40, 41, 140; museum 41–2; theory 40
posthumous 4, 18, 140; theory 17, 123; philosophy 35
post-preservation 62
Povinelli, Elizabeth 11, 63, 115

Rachel Carson Center for Environment and Society 37
Rancièr, Jacques 53
Rinehart, Gina 18
Rio Tinto 9, 16, 17, 96
Rocky Mountain National Park 102
Rowell, Steve: *Uncanny Sensing, Remote Valleys* 16–17
Royal Society, the 80, 83, 121
ruin studies 28, 62
Russell, Andrew and Lee Vinsel 49
Rutherford-Morrison, Laura 85

Sacred Stone Camp 16
Saldanha, Arun 76
Samarco dam disaster 50
Saraceno, Tomás 13
scale critique 31
Science of Survival, The – Your Planet Needs You 42
shadow governments 3, 96
shadow places 21
Sheldrake, Merlin 37; *Entangled Life* (book) 34
Shell 17
Shiva, Vandana 26
Simon, Nina 52
singularity 41
Skelleftea Museum 53
Smithsonian Institutions, Washington DC 16, 25
soils 18; and art 20, 22–5; and climate 20, 22; compost 23; and death 23; *Deep Roots*; *Dig It! The Secrets of Soil* 25; disappearance of 3, 6, 20; kinship with 1; and language 32; leachate; microbiomes 22, 31; as monoliths 22, 118; pedometry 20, 21; and rubble 27; as symbols 32; SSAA 25; as underland 25, 33; World Soil Museum 22; United Nations 2015 Year of Soils 32
Soil Science Association America (SSSA) 25
solastalgia 14, 95
Sontag, Susan *The Volcano Lover* 81
Sovereign Hill living history museum 125, 129
spacetimemattering 57
Sparrow, Jeff 43, 88, 138
speculative realism 17
Spindletop-Gladys City Boomtown Museum 109

Standing Rock Sioux camp 63
Stanley Robinson, Kim 136; *2312* 64; *The Ministry for the Future* 64
Steffen, Will 69
Stephens, Beth and Annie Sprinkle, ecosexual soil weddings 24
Stoppani, Antonio 67, 90
Sterling, Colin 4
storytelling 21, 23, 24, 26, 31; climate fiction 64, 64; petrofiction 100; stone lore 62; of world-making 66
strata 83; and stratigraphy 68, 114
stump-jump plough 21
sublime underworlds 33
Svalbard Global Seed Vault 26, 122

Tate Britain 47–8, 103
Tate Modern 103
techno-utopia 7, 13, 18, 40; as discourse 3, 76; and museums 3
Terrestrial, the 14, 33, 73, 76, 77
Thin Ice 103–4
Thomas the Tank Engine 91
Þórsson, Bergsveinn 39
Toadvine, Ted 64, 75, 76
Todd, Zoe 15
Tøndborg, Britta 53
trans-humanism: 3, 18, 43, 52, 84, 105, 140; critique of 40–1; Ray Kurzweil and Hans Moravec 41; World Transhumanist Association 41
Tsing, Anna *Arts of Living on a Damaged Planet* (book) 12
Tuvalu 38, 45

Ukraine, 2022 invasion of 10, 89
Uncanny Sensing, Remote Valleys 16–17
underland 24, 28, 34; *see also* Macfarlane, Robert
United Nations 2015 Year of Soils 33
universal museum, the 52, 58, 70; Declaration on the Importance and Value of Universal Museums 52

universities 7, 75, 76, 103; closure of humanities courses 49; and innovation rhetoric 48; and managerialism 48
unweather 59

van Alphen, Ernst 98
van Dooren, Thom 26
Verne, Jules 112
vibrant matter 18
volcanoes 71, 81, 118, 135
von Humboldt, Alexander 113
V. V. Dokuchaev Central Museum of Pedology 22

Water Will Be Here 47
Welcome to the Anthropocene: The Earth in our Hands exhibition 37–9, 94
Werner, Abraham Gottlob 81
Western Australian Museum of the Goldfields 125, 128
Western Australian School of Mines 60
White Geology, the 3, 18; history of 83–5, 126, 141; power of 78, 81, 86, 88; and space junk 84
Wiess Energy Hall 118
Williams, Raymond 15, 68
Wilson, Fred 16
Winfrey, Oprah 58
Wolfe, Cary 40, 48, 52
Woodward, John 81
world, the 15, 33, 52, 75; end of 142
World Transhumanist Association 41
World Soil Museum, Wageningen University 22
Wray, Lynn 53–4

Yanacocha Gold Mine 132
Young, Emily 114
Yucca Mountain, Nevada 29
Yusoff, Kathryn 3, 61, 78

Zalasiewicz, Jan 31
Žižek, Slavoj 53, 65
Zone à Défedre (ZAD) 13
Zurkow, Marina: Wink, Texas 105–6

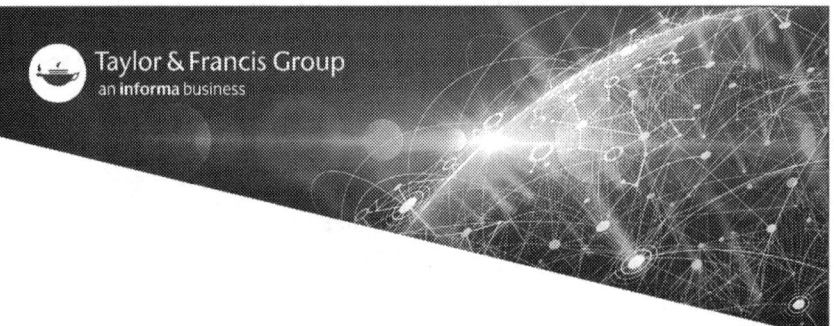

Printed in the United States
by Baker & Taylor Publisher Services